Advanced Ceramics and Composites

Volume 2

Series Editor

Longbiao Li, College of Civil Aviation, Nanjing University of Aeronautics and Astronautics, Nanjing, Jiangsu, China

The book series "Advanced Ceramics and Composites" publishes insights and latest research results on advanced ceramics and composites, as well as the applications of these materials. The intent is to cover all the technical contents, applications, and multidisciplinary aspects of Advanced Ceramics and Composites.

The objective of the book series is to publish monographs, reference works, selected contributions from specialized conferences, and textbooks with high quality in the field advanced ceramics and composite materials. The series provides valuable references to a wide audience in research community in materials science, research and development personnel of ceramic and composite materials, industry practitioners and anyone else who are looking to expand their knowledge of ceramics and composites.

More information about this series at http://www.springer.com/series/16543

Longbiao Li

Damage and Fracture of Ceramic-Matrix Composites Under Stochastic Loading

Springer

Longbiao Li
College of Civil Aviation
Nanjing University of Aeronautics and Astronautics
Nanjing, Jiangsu, China

ISSN 2662-9305 ISSN 2662-9313 (electronic)
Advanced Ceramics and Composites
ISBN 978-981-16-2140-6 ISBN 978-981-16-2141-3 (eBook)
https://doi.org/10.1007/978-981-16-2141-3

This Springer imprint is published by the registered company Springer Nature Singapore Pte Ltd.
The registered company address is: 152 Beach Road, #21-01/04 Gateway East, Singapore 189721, Singapore

To My Son Shengning Li

Preface

Ceramic-matrix composites (CMCs) possess low material density (i.e., only 1/4–1/3 of high-temperature alloy) and high temperature resistance, which can reduce cooling air and improve structure efficiency. Understanding the failure mechanisms and internal damage evolution represents an important step to ensure reliability and safety of CMCs. This book focuses on the damage and fracture of CMCs under stochastic loading, as following:

- Effect of stochastic loading on tensile damage and fracture of fiber-reinforced CMCs is investigated. A micromechanical constitutive model is developed considering multiple damage mechanisms under tensile loading. Relationship between stochastic stress, tangent modulus, interface debonding, and fiber broken is established. Effects of fiber volume, interface shear stress, interface debonding energy, saturation matrix crack spacing, and fiber strength on tensile stress–strain curve, tangent modulus, interface debonding fraction, and fiber broken fraction are analyzed. Experimental tensile damage and fracture of unidirectional and 2D SiC/SiC composites subjected to different stochastic loading stress are predicted.
- Synergistic effects of stochastic loading sequence and interface wear on fatigue hysteresis loops of fiber-reinforced CMCs are investigated. Based on the fatigue damage mechanism of fiber sliding relative to matrix in the interface debonding region, the interface debonding length, unloading interface counter-slip length, and reloading interface new-slip length are determined using fracture mechanic approach. Effects of stochastic stress level, material properties, interface wear, and stochastic loading sequence on the interface slip and fatigue hysteresis loops are analyzed.
- Stress-rupture of fiber-reinforced CMCs with stochastic loading at intermediate temperature is investigated. Four different stochastic loading sequences under stress-rupture are considered with different loading time and time spacing. Microstress field of the damaged CMCs, damage models of matrix cracking, interface debonding, and fiber failure consider stochastic loading and time-dependent interface and fiber oxidation. Damage evolution of stress-rupture strain, interface debonding and interface oxidation ratio, and the broken fibers fraction versus time curves of SiC/SiC composite subjected to four different stochastic loading

sequences are analyzed. Effects of fiber volume, matrix crack spacing, interface debonding energy, interface shear stress, and environmental temperature on damage evolution and lifetime of SiC/SiC composite are discussed. Experimental damage evolution and lifetime of SiC/SiC composite under stress-rupture with stochastic loading are predicted.

- A micromechanical approach is developed to predict fatigue life of fiber-reinforced CMCs subjected to stochastic overloading stress at elevated temperature. Multiple elevated temperature fatigue damage mechanisms of interface wear and interface and fiber oxidation are considered in the analysis of elevated temperature fatigue damage evolution and fracture under stochastic overloading. Relationships between stochastic overloading stress level and occurrence applied cycle number, broken fibers fraction, and fatigue life decreasing rate are established. Experimental fatigue life S–N curves of C/SiC composites with different fiber preforms (i.e., unidirectional, cross-ply, 2D woven, 2.5D woven, and 3D braided) are predicted for different stochastic overloading stress levels and occurrence applied cycle number.
- A micromechanical fatigue limit stress model of fiber-reinforced CMCs subjected to stochastic overloading stress is developed. Fatigue limit stress of different C/SiC composites (i.e., unidirectional, cross-ply, 2D, 2.5D, and 3D C/SiC) is predicted based on the micromechanical fatigue damage models and fatigue failure criterion. Under cyclic fatigue loading, the fatigue damage and fracture under stochastic overloading stress at different applied cycle number are characterized using two parameters of fatigue life decreasing rate and broken fiber fraction. Relationships between fatigue life decreasing rate, stochastic overloading stress and corresponding occurrence applied cycle number, and broken fibers fraction are analyzed.

I hope this book can help the material scientists and engineering designers to understand and master the damage and fracture of ceramic-matrix composites under stochastic loading.

Nanjing, China
January 2021

Longbiao Li

Acknowledgements I am grateful to my mum Yuping Zhou, my wife Li Peng, and my son Shengning Li for their encouragement.

A special thanks to Dr. Mengchu Huang and Chandra Sekaran Arjunan for their help with my original manuscript.

I am also grateful to the team at Springer for their professional assistance.

Contents

Chapter 1
Tensile Damage and Fracture of Ceramic-Matrix Composites Subjected to Stochastic Loading

Abstract In this chapter, effect of stochastic loading on tensile damage and fracture of fiber-reinforced ceramic-matrix composites (CMCs) is investigated. A micromechanical constitutive model is developed considering multiple damage mechanisms under stochastic tensile loading. Relationships between stochastic stress, composite tangent modulus, interface debonding, and fibers broken are established. Effects of composite constitutive properties (i.e., fiber volume, interface shear stress, interface debonding energy, and fiber strength) and composite damage state (i.e., saturation matrix crack spacing) on the tensile stress–strain curve, composite tangent modulus, interface debonding ratio, and fibers broken fraction are analyzed. Experimental tensile damage and fracture of unidirectional and 2D SiC/SiC composites subjected to different stochastic loading stresses are predicted.

Keywords Ceramic-matrix composites (CMCs) · Stochastic loading · Tensile · Damage · Fracture

1.1 Introduction

SiC ceramic has the advantages of high temperature resistance, wear resistance, oxidation resistance, and high specific strength, but the impact resistance of monolithic ceramics is poor. After toughening with continuous SiC fiber, the SiC/SiC ceramic-matrix composites (CMCs) possess good thermal shock resistance, insensitive to cracks, and no catastrophic damage on the basis of retaining the ceramic properties (Naslain 2004; Li et al. 2020). Compared with the superalloy, SiC_f/SiC composite possesses lower density (i.e., 2.0−3.0 g/cm^3, only 1/3−1/4 of that of superalloy) and higher temperature resistance (>1200 °C under noncooling conditions). It can improve the turbine inlet temperature (TIT), reduce the amount of cooling air, improve the combustion efficiency, reduce the structural quality, so as to improve the thrust-to-weight ratio (TWR) of the aeroengine, and become an ideal hot-section component material for high-performance aero engine (Ruggles-Wrenn et al. 2011; Li t al. 2014; Liu 2019; Guo and Kagawa 2002; Morscher 1999; Li 2020).

After decades of development, a variety of SiC/SiC composite preparation methods have been developed, such as chemical vapor infiltration (CVI)

L. Li, *Damage and Fracture of Ceramic-Matrix Composites Under Stochastic Loading*, Advanced Ceramics and Composites 2,
https://doi.org/10.1007/978-981-16-2141-3_1

(Udayakumar et al. 2011), polymer impregnation pyrolysis (PIP) (Yin et al. 2015), and melt infiltration (MI) (Corman and Luthra 2004), etc. The main difference between the three processes lies in the way for densification of the SiC matrix. In CVI process, the gaseous precursor is pyrolysis and deposited on the surface of SiC fiber to obtain the SiC matrix; PIP process usually impregnates the fiber preform in liquid precursor, then the precursor is vitrified by pyrolysis at high temperature, and the dense SiC matrix is obtained by repeated immersion pyrolysis process; MI process is to infiltrate molten silicon into porous carbon preform, and the carbon reacts with silicon to form the SiC matrix. The properties of SiC_f/SiC composites are closely related to the fabrication process, and there are some differences in the properties of materials prepared by different processes (Liu et al. 2020). General Electric (GE) company combines the prepreg technology with the melt infiltration process and develops the prepreg melt infiltration (Prepreg-MI) process with the advantages of short preparation cycle, low cost, and strong designability. SiC/SiC composites with high density, good reliability, and excellent comprehensive performance are prepared by using this process, which has already been successfully applied to the turbine outer ring of aeroengine and will be further applied to the hot-section components of aeroengine, such as combustion liner, guide vane, and blade, etc (Corman and Luthra 2004).

In-plane tensile properties reflect the strength of the composite's resistance to external tensile load (Morscher et al. 2007). Under monotonic tensile loading, multiple damage mechanisms occur in the matrix, interphase, and fibers. Based on the microdamage observation and analysis, tensile stress–strain curves can be divided into four stages, including:

- Stage I, linear-elastic region.
- Stage II, matrix cracking and interface debonding region. Matrix microcracking occurs first and the fiber debonding from the matrix, leading to nonlinear behavior of CMCs.
- Stage III, saturation matrix cracking region. Matrix cracking approaches saturation with complete debonding of the fiber from the matrix.
- Stage IV, fiber failure region. The fiber gradually fractures with increasing applied stress.

The composite elastic modulus, proportional limit stress, ultimate tensile strength, and the corresponding fracture strain can be obtained from the tensile stress–strain curves. For the SiC/SiC composite fabricated using the MI method, the fracture strength and failure strain of Hi-Nicalon™ Type S SiC/SiC composite are much higher than that of Tyranno™ SA3 SiC/SiC composite; however, the initial elastic modulus and proportional limit stress of Hi-Nicalon™ Type S SiC/SiC composite are lower than that of Tyranno™ SA3 SiC/SiC composite at the same fiber volume $V_f = 34.8\%$; and the composite tensile strength increases with the fiber volume (Morscher and Pujar 2009). Marshall et al. (1985), Zok and Spearing (1992) investigated the first matrix cracking and matrix multiple cracking evolution in fiber-reinforced CMCs using the fracture mechanics approach. Curtin (1993) investigated multiple matrix cracking evolution of CMCs considering matrix internal flaws. Evans (1997) developed an approach for design and life prediction issues for fiber-reinforced CMCs and established the relationship between macromechanical behavior and constituent properties of CMCs. McNulty and Zok (1999) investigated the low-cycle fatigue damage mechanism and established the damage models for predicting the low-cycle fatigue life of CMCs. Naslain et al. (1999) investigated the monotonic and cyclic tensile behavior of Nicalon™ SiC/SiC minicomposite at room temperature. The relationship between the statistical parameters of both the fiber and the matrix, the fiber/matrix interfacial parameters, the effect of environment, and the tensile curves has been established. Goto and Kagawa (1996) investigated the tensile and fracture behavior of a bidirectional woven Nicalon™ SiC/SiC composite at room temperature. The tensile stress–strain curve shows nonlinear behavior above $\sigma = 70$ MPa, and the composite effective Young's modulus at fracture was approximately 40% of the initial value. Guo and Kagawa (2002) investigated the tensile fracture behavior and tensile mechanical properties of Nicalon™ and Hi-Nicalon™ SiC/BN/SiC composites at temperature between $T = 298$ and 1400 K in air atmosphere. The tensile strength dropped from $\sigma_{\text{uts}} = 140$ MPa at $T = 800$ K to $\sigma_{\text{uts}} = 41$ MPa at $T = 1200$ K. Meyer and Waas (2018) investigated tensile response of notched SiC/SiC composite at elevated temperature using a novel digital image correlation technique. Li et al. (2013, 2015; Li 2018) developed a micromechanical approach to predict the tensile behavior of unidirectional, cross-ply, 2D woven, and 2.5D woven fiber-reinforced CMCs considering the damage mechanisms in the matrix, interphase, and fibers. However, during real applications of aeroengine components, stochastic loading appears and affects the tensile mechanical behavior of fiber-reinforced CMCs. It is necessary to perform investigation on the effect of stochastic loading on tensile damage and fracture of fiber-reinforced CMCs.

In this chapter, effect of stochastic loading on tensile damage and fracture of fiber-reinforced CMCs is investigated. A micromechanical constitutive model considering stochastic loading is developed considering multiple damage mechanisms under tensile loading. Relationships between stochastic loading stress, composite tangent modulus, interface debonding, and fibers broken are established. Effects of composite constituent properties and damage stage on composite tensile stress–strain curve, composite tangent modulus, interface debonding ratio, and fibers broken fraction are analyzed. Experimental tensile damage and fracture of unidirectional and 2D SiC/SiC composites subjected to different stochastic loading stress are predicted.

1.2 Theoretical Model

Under stochastic loading, serious damage mechanisms and modes occur inside of CMCs and affect the tensile mechanical response of the composites. Figure 1.1 shows multiple stochastic loading sequence under tensile loading. The shear-lag model is used to analyze microstress field of damaged composite. A unit cell is extracted from

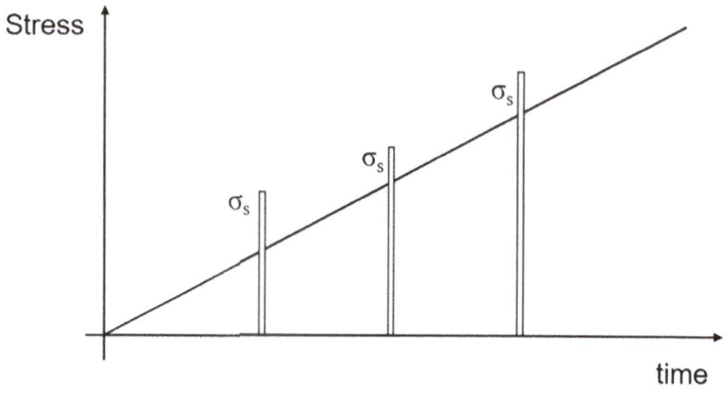

Fig. 1.1 Diagram of stochastic loading under tensile loading

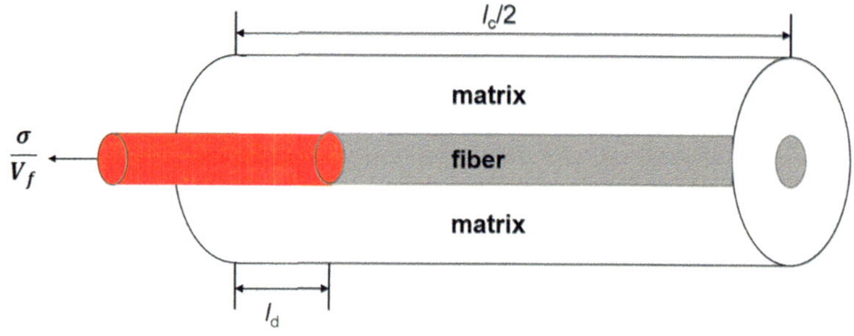

Fig. 1.2 Unit cell of damaged composite

the ceramic composite system, as shown in Fig. 1.2. The fiber axial stress in different damage regions is given by:

$$\sigma_{\mathrm{f}}(x) = \begin{cases} \Phi_{\mathrm{s}} - \dfrac{2\tau_{\mathrm{i}}}{r_{\mathrm{f}}}x, x \in [0, l_{\mathrm{d}}] \\ \sigma_{\mathrm{fo}} + \left(\Phi_{\mathrm{s}} - \sigma_{\mathrm{fo}} - 2\dfrac{l_{\mathrm{d}}}{r_{\mathrm{f}}}\tau_{\mathrm{i}}\right)\exp\left(-\rho\dfrac{x - l_{\mathrm{d}}}{r_{\mathrm{f}}}\right), x \in \left[l_{\mathrm{d}}, \dfrac{l_{\mathrm{c}}}{2}\right] \end{cases} \tag{1.1}$$

where Φ_{s} is the intact fiber stress under stochastic stress σ_{s}, τ_{i} is the interface shear stress, r_{f} is the fiber radius, σ_{fo} is the fiber axial stress in the interface bonding region, l_{d} is the interface debonding length, l_{c} is the matrix crack spacing, and ρ is the shear-lag model parameter.

Using stochastic matrix cracking model, relationship between stochastic stress and matrix crack spacing is given by (Curtin 1993):

$$l_{\mathrm{c}} = l_{\mathrm{s}}\left\{1 - \exp\left[-\left(\frac{\sigma_{\mathrm{m}}}{\sigma_{\mathrm{R}}}\right)^{m}\right]\right\}^{-1} \tag{1.2}$$

where l_s is the saturation matrix crack spacing, σ_m is the matrix stress, σ_R is matrix cracking characteristic stress, and m is the matrix Weibull modulus.

The fracture mechanics approach is used to determine the interface debonding length.

$$l_d = \frac{r_f}{2}\left(\frac{V_m E_m \sigma_s}{V_f E_c \tau_i} - \frac{1}{\rho}\right) - \sqrt{\left(\frac{r_f}{2\rho}\right)^2 + \frac{r_f V_m E_m E_f}{E_c \tau_i^2}\Gamma_d} \tag{1.3}$$

where V_f and V_m are the fiber and matrix volume fraction, respectively, E_m and E_c are the matrix and composite elastic modulus, respectively, and Γ_d is the interface debonding energy.

The global load sharing (GLS) criterion is used to determine the intact fiber stress under stochastic loading stress. Relationship between the fiber intact stress, fiber failure probability, and fiber pullout length is given by (Curtin 1991):

$$\frac{\sigma_s}{V_f} = \Phi(1 - q(\Phi_s)) + \frac{2\tau_i}{r_f}\langle L\rangle q(\Phi) \tag{1.4}$$

where $\langle L\rangle$ is the average fiber pullout length and $q(\Phi_s)$ is the fiber failure probability.

$$q(\Phi_s) = 1 - \exp\left[-\left(\frac{\Phi_s}{\sigma_{fc}}\right)^{m_f+1}\right] \tag{1.5}$$

where σ_{fc} is the fiber characteristic strength and m_f is the fiber Weibull modulus.

When matrix cracking and interface debonding occur, the composite strain is

$$\varepsilon_c = \begin{cases} \frac{\sigma}{V_f E_f}\eta - \frac{\tau_i}{E_f}\frac{l_d}{r_f}\eta + \frac{\sigma_{fo}}{E_f}(1-\eta) - \frac{2}{\rho E_f}\left(\frac{V_m}{V_f}\frac{r_f}{l_c}\sigma_{mo} - \eta\tau_i\right) \\ \quad \times \left[\exp\left(-\frac{\rho}{2}\frac{l_c}{r_f}(1-\eta)\right) - 1\right] - (\alpha_c - \alpha_f)\Delta T, \eta < 1 \\ \frac{\sigma}{V_f E_f} - \frac{1}{2}\frac{\tau_i}{E_f}\frac{l_c}{r_f} - (\alpha_c - \alpha_f)\Delta T, \eta = 1 \end{cases} \tag{1.6}$$

When fibers failure occurs, the composite strain is

$$\varepsilon_c = \begin{cases} \frac{\Phi}{E_f}\eta - \frac{\tau_i}{E_f}\frac{l_d}{r_f}\eta + \frac{\sigma_{fo}}{E_f}(1-\eta) - \frac{2}{\rho E_f}\left(\frac{r_f}{l_c}(\Phi - \sigma_{fo}) - \eta\tau_i\right) \\ \quad \times \left[\exp\left(-\frac{\rho}{2}\frac{l_c}{r_f}(1-\eta)\right) - 1\right] - (\alpha_c - \alpha_f)\Delta T, \eta < 1 \\ \frac{\Phi}{E_f} - \frac{1}{2}\frac{\tau_i}{E_f}\frac{l_c}{r_f} - (\alpha_c - \alpha_f)\Delta T, \eta = 1 \end{cases} \tag{1.7}$$

where η is the interface debonding ratio.

$$\eta = 2\frac{l_d}{l_c} \tag{1.8}$$

The composite tangent modulus is defined as

$$E_p = \frac{d\sigma}{d\varepsilon} \tag{1.9}$$

1.3 Results and Discussion

Under stochastic loading, multiple damage mechanisms, and modes of the matrix cracking, interface debonding and fiber failure occur. The microstress field of the damaged CMCs after stochastic loading is given by Eq. (1.1). The fiber axial stress distribution is affected by the stochastic loading stress level, matrix cracking, interface debonding, and fiber failure. The stochastic matrix cracking model is used to determine the matrix crack spacing at the applied stress level, as shown in Eq. (1.2), and the fracture mechanics interface debonding criterion is used to determine the interface debonding length, and the GLS criterion is used to determine the load allocation between fracture and intact fibers. The micromechanical constitutive models for the conditions of the matrix cracking, interface debonding, and fiber failure are given by Eqs. (1.6) and (1.7). Using the developed micromechanical constitutive models and damage models, the effects of the fiber volume, interface shear stress, interface debonding energy, matrix crack spacing, and fiber strength on tensile stress–strain curve, tangent modulus, interface debonding fraction, and fiber broken fraction of SiC/SiC composite subjected to different stochastic loading are analyzed. Li et al. (2014) investigated the tensile behavior of 2D woven SiC/SiC composite at room temperature. The composite was fabricated using the CVI method with the PyC interphase. The tensile experiments were conducted under displacement control of 0.3 mm/min. The material properties are given by: $V_f = 0.2$, $E_f = 350$ GPa, $E_m = 300$ GPa, $r_f = 7.5$ μm, $\alpha_f = 4.0 \times 10^{-6}$/°C, $\alpha_m = 4.8 \times 10^{-6}$/°C, $\Delta T = -1000$ °C, $\Gamma_d = 0.5$ J/m^2, $\tau_i = 25$ MPa, and $m_f = 3$.

1.3.1 Tensile Damage and Fracture of SiC/SiC Composite with Stochastic Loading for Different Fiber Volumes

The fiber volume affects the tensile behavior of CMCs. The fiber volume range of SiC/SiC composite is between $V_f = 27.7$ and 40% (Li 2020a). In the present analysis, the effect of the fiber volume (i.e., $V_f = 0.3$ and 0.35) on the composite tensile stress–strain curves, composite tangent modulus, interface debonding ratio, and broken fibers fraction of SiC/SiC composite subjected to different stochastic loading stress $\sigma_s = 200$, 300, and 350 MPa are shown in Figs. 1.3 and 1.4 and Table 1.1. When the fiber volume increases, the stress carried by the fiber increases, the initial composite strain decreases, the initial composite tangent modulus increases, the transition stress for the interface debonding decreases, and the fibers broken fraction at initial loading decreases.

For the fiber volume of $V_f = 0.3$, the composite damage mechanisms and tensile damage evolution and fracture subjected to different stochastic loading stresses of $\sigma_s = 200$, 300, and 350 MPa are analyzed, as following:

- Under $\sigma_s = 200$ MPa, multiple damage modes of matrix cracking and interface debonding occur, leading to the increase of the composite initial strain, decreasing of the composite tangent modulus, and increase of the broken fibers fraction. The initial composite strain is $\varepsilon_0 = 0.00161\%$ due to the matrix cracking and interface debonding at stochastic loading stress $\sigma_s = 200$ MPa; the composite initial tangent modulus is approximately $E_p = 268$ GPa, and the degradation rate of the composite tangent modulus is 15% compared with the original specimen, and the fibers broken fraction is $q = 0.005$. With increasing stress to $\sigma_{tr} = 79.2$ MPa, the interface debonding ratio increases, the composite tangent modulus decreases to approximately $E_p = 235$ GPa corresponding to the interface debonding ratio $\eta = 0.081$. Upon increasing stress from $\sigma_{tr} = 79.2$ MPa to $\sigma = 200$ MPa, the composite tangent modulus remains constant value of approximately $E_p = 264$ GPa with the interface debonding ratio $\eta = 0.084$.
- Under $\sigma_s = 300$ MPa, the initial composite strain is $\varepsilon_0 = 0.00982\%$ due to the damage modes of the matrix cracking and interface debonding; the composite initial tangent modulus is approximately $E_p = 229$ GPa, and the degradation rate of the composite tangent modulus is 28% compared with the original specimen, and the fibers broken fraction is $q = 0.028$. With increasing applied stress to $\sigma_{tr} = 178.2$ MPa, the interface debonding fraction increases, the composite tangent modulus decreases to approximately $E_p = 145$ GPa corresponding to the interface debonding ratio of $\eta = 0.392$. Upon increasing applied stress from $\sigma_{tr} = 178.2$ MPa to $\sigma = 300$ MPa, the composite tangent modulus remains constant value of approximately $E_p = 170.1$ GPa with the interface debonding ratio of $\eta = 0.4$.
- Under $\sigma_s = 350$ MPa, the initial composite strain is $\varepsilon_0 = 0.023\%$ due to the damages of the matrix cracking and interface debonding; the composite initial tangent modulus is approximately $E_p = 228$ GPa, and the degradation rate of composite tangent modulus is 28% compared with the original specimen, and the fibers broken fraction is $q = 0.06$. With increasing applied stress to $\sigma_{tr} = 231$ MPa, the interface debonding fraction increases, the composite tangent modulus decreases to approximately $E_p = 131$ GPa corresponding to the interface debonding ratio of $\eta = 0.51$. Upon increasing applied stress from $\sigma_{tr} = 131$ MPa to $\sigma = 350$ MPa, the composite tangent modulus remains constant value of approximately $E_p = 151$ GPa with the interface debonding ratio of $\eta = 0.51$.

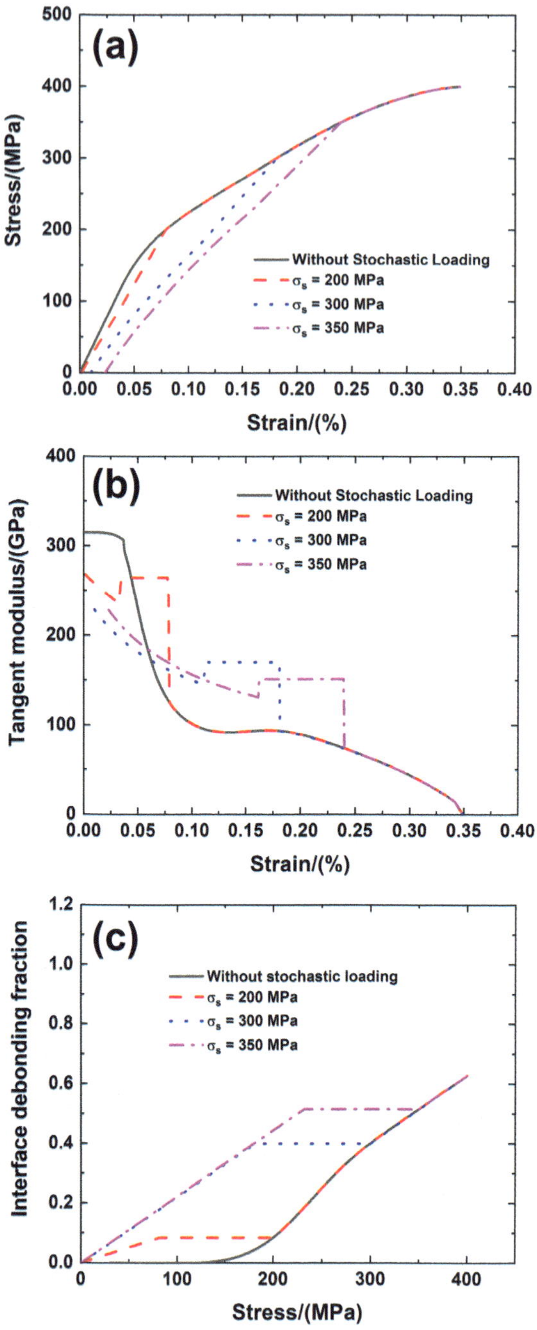

Fig. 1.3 **a** Tensile stress–strain curves; **b** the composite tangent modulus versus applied strain curves; **c** the interface debonding ratio versus applied stress curves; and **d** the broken fibers fraction versus applied stress curves of SiC/SiC composite under $\sigma_s = 200$, 300, and 350 MPa when $V_f = 0.3$

Fig. 1.3 (continued)

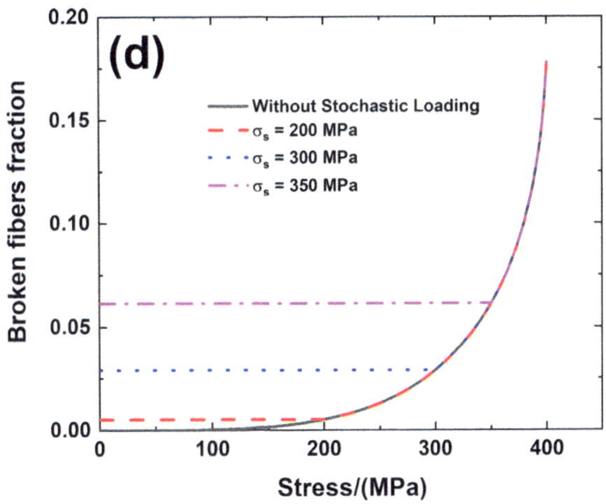

For the fiber volume of $V_f = 0.3$, the composite damage mechanisms and tensile damage evolution and fracture subjected to different stochastic loading stresses of $\sigma_s = 200$, 300, and 350 MPa are analyzed, as following:

- Under $\sigma_s = 200$ MPa, the damage modes of matrix cracking and interface debonding occur, leading to the increase of the composite initial strain, decreasing of the composite tangent modulus, and increase of the broken fibers fraction. The initial composite strain is $\varepsilon_0 = 0.0009\%$; the composite initial tangent modulus is approximately $E_p = 281$ GPa, the degradation rate of composite tangent modulus is 12% compared with original specimen, and the fibers broken fraction is $q = 0.0027$. With increasing applied stress to $\sigma_{tr} = 55$ MPa, the interface debonding ratio increases and the composite tangent modulus decreases to approximately $E_p = 265$ GPa corresponding to the interface debonding ratio of $\eta = 0.046$. Upon increasing applied stress from $\sigma_{tr} = 55$ MPa to $\sigma = 200$ MPa, the composite tangent modulus remains constant value of approximately $E_p = 291$ GPa with the interface debonding ratio of $\eta = 0.046$.
- Under $\sigma_s = 300$ MPa, the initial composite strain is $\varepsilon_0 = 0.0047\%$ due to the damage modes of matrix cracking and interface debonding; the composite initial tangent modulus is approximately $E_p = 249$ GPa, and the degradation rate of composite tangent modulus is 22% compared with original specimen, and the fibers broken fraction is $q = 0.014$. With increasing applied stress to $\sigma_{tr} = 156.2$ MPa, the interface debonding ratio increases, and the composite tangent modulus decreases to approximately $E_p = 201$ GPa corresponding to the interface debonding ratio of $\eta = 0.27$. When the applied stress increases from $\sigma_{tr} = 156.2$ MPa to $\sigma = 250$ MPa, the composite tangent modulus remains constant value of approximately $E_p = 216$ GPa with the interface debonding ratio of $\eta = 0.27$.
- Under $\sigma_s = 350$ MPa, the initial composite strain is $\varepsilon_0 = 0.0095\%$ due to the damage modes of the matrix cracking and interface debonding; the composite

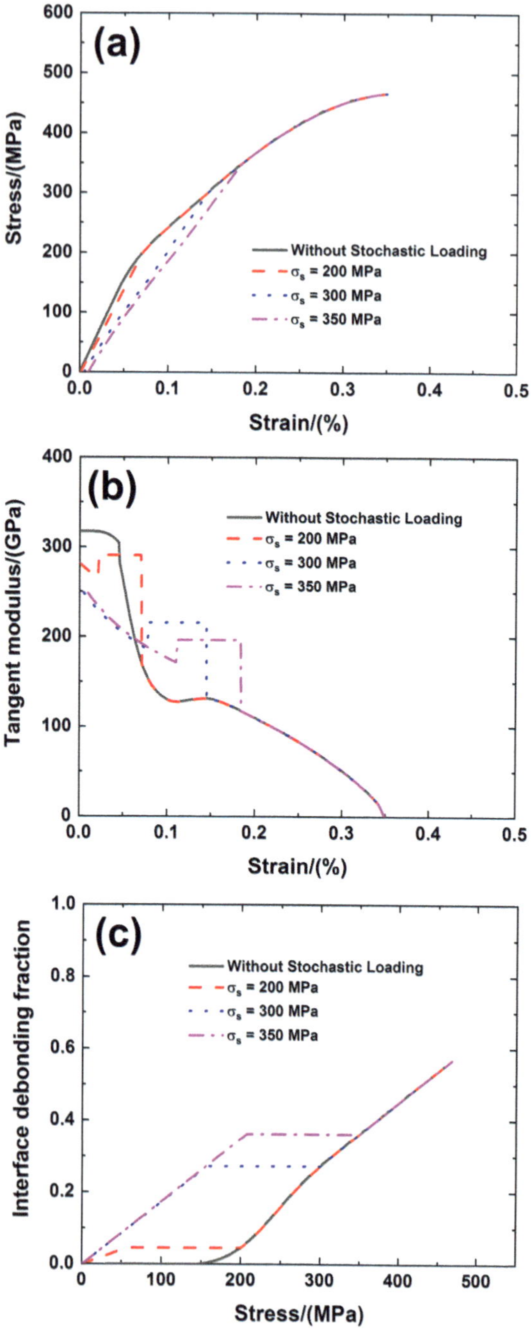

Fig. 1.4 **a** Tensile stress–strain curves; **b** the composite tangent modulus versus applied strain curves; **c** the interface debonding ratio versus applied stress curves; and **d** the broken fibers fraction versus applied stress curves of SiC/SiC composite under $\sigma_s = 200$, 300, and 350 MPa when $V_f = 0.35$

Fig. 1.4 (continued)

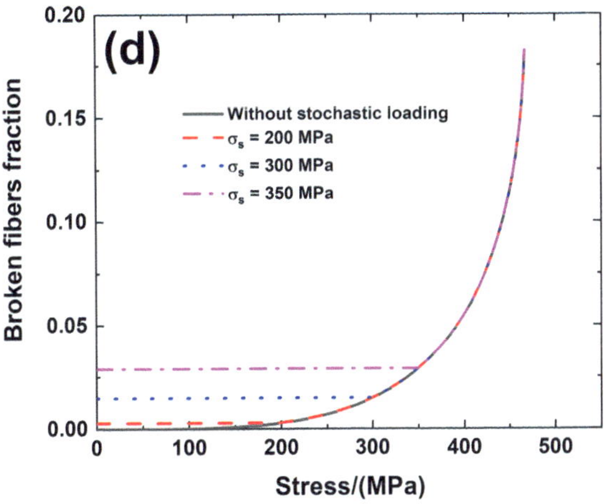

Table 1.1 Effect of the fiber volume ($V_f = 0.3$ and 0.35) on tensile damage parameters of SiC/SiC composite subjected to $\sigma_s = 200$, 300, and 350 MPa

$V_f = 0.3$	$\sigma_s = 200$ MPa			
	$\varepsilon_0/(\%)$	$E_p/(\text{GPa})$	$\sigma_{tr}/(\text{MPa})$	$q/(\%)$
	0.00161	268	79.2	0.005
	$\sigma_s = 300$ MPa			
	$\varepsilon_0/(\%)$	$E_p/(\text{GPa})$	$\sigma_{tr}/(\text{MPa})$	$q/(\%)$
	0.0098	229	178.2	0.028
	$\sigma_s = 350$ MPa			
	$\varepsilon_0/(\%)$	$E_p/(\text{GPa})$	$\sigma_{tr}/(\text{MPa})$	$q/(\%)$
	0.023	228	231	0.060
$V_f = 0.35$	$\sigma_s = 200$ MPa			
	$\varepsilon_0/(\%)$	$E_p/(\text{GPa})$	$\sigma_{tr}/(\text{MPa})$	$q/(\%)$
	0.0009	281	55	0.0027
	$\sigma_s = 300$ MPa			
	$\varepsilon_0/(\%)$	$E_p/(\text{GPa})$	$\sigma_{tr}/(\text{MPa})$	$q/(\%)$
	0.0047	249	156.2	0.014
	$\sigma_s = 350$ MPa			
	$\varepsilon_0/(\%)$	$E_p/(\text{GPa})$	$\sigma_{tr}/(\text{MPa})$	$q/(\%)$
	0.0095	248	206.8	0.028

initial tangent modulus is approximately $E_p = 248$ GPa, the degradation rate of composite tangent modulus is 22% compared with original specimen, and the fibers broken fraction is $q = 0.028$. With increasing applied stress to $\sigma_{tr} = 206.8$ MPa, the interface debonding ratio increases and the composite tangent modulus decreases to approximately $E_p = 191.9$ GPa corresponding to the interface debonding ratio of $\eta = 0.362$. When the applied stress increases from $\sigma_{tr} = 206.8$ MPa to $\sigma = 300$ MPa, the composite tangent modulus remains constant value of $E_p = 196.7$ GPa with the interface debonding ratio of $\eta = 0.363$.

1.3.2 Tensile Damage and Fracture of SiC/SiC Composite with Stochastic Loading for Different Interface Shear Stress

The interface shear stress transfers the load between the fiber and the matrix when the damage modes of matrix cracking and interface debonding occur. For the weak interface bonding between the fiber and the matrix of SiC/SiC composite, the value of the interface shear stress is between the range of $\tau_i = 10$ and 30 MPa (Li 2014). In the present analysis, the effect of the interface shear stress (i.e., $\tau_i = 15$, and 20 MPa) on the tensile damage and fracture of SiC/SiC composite subjected to $\sigma_s = 200$, 230, and 250 MPa are analyzed, as shown in Figs. 1.5 and 1.6 and Table 1.2. When the interface shear stress increases, the load transfer capacity between the fiber and the matrix increases, the initial composite strain remains the same, the initial tangent modulus increases, the transition stress for interface debonding remains the same, and the initial fiber broken fraction remains the same.

When $\tau_i = 15$ MPa, the composite damage mechanisms and tensile damage evolution and fracture subjected to different stochastic loading stresses of $\sigma_s = 200$, 230, and 250 MPa are analyzed, as following:

- Under $\sigma_s = 200$ MPa, the damage modes of matrix cracking and interface debonding occur, leading to the increase of the composite initial strain, decreasing of the composite tangent modulus, and increase of the broken fibers fraction. The initial composite strain is $\varepsilon_0 = 0.00943\%$; the composite initial tangent modulus is approximately $E_p = 195.6$ GPa, the degradation rate of composite tangent modulus is 37% compared with original specimen, and the fibers broken fraction is $q = 0.028$. With increasing applied stress to $\sigma_{tr} = 125.4$ MPa, the interface debonding ratio increases, the composite tangent modulus decreases to approximately $E_p = 108.1$ GPa corresponding to the interface debonding ratio of $\eta = 0.374$. When the stress increases from $\sigma_{tr} = 125.4$ MPa to $\sigma = 200$ MPa, the composite tangent modulus remains constant value of $E_p = 131.3$ GPa with the interface debonding ratio of $\eta = 0.382$.
- Under $\sigma_s = 230$ MPa, the initial composite strain is $\varepsilon_0 = 0.021\%$ due to the damage modes of matrix cracking and interface debonding; the composite initial tangent modulus is approximately $E_p = 163$ GPa, the degradation rate of composite tangent modulus is 48% compared with original specimen, and the fibers broken fraction is $q = 0.056$. With increasing applied stress to $\sigma_{tr} = 154$ MPa, the interface debonding ratio increases, and the composite tangent modulus decreases to approximately $E_p = 71.5$ GPa corresponding to the interface debonding ratio of η

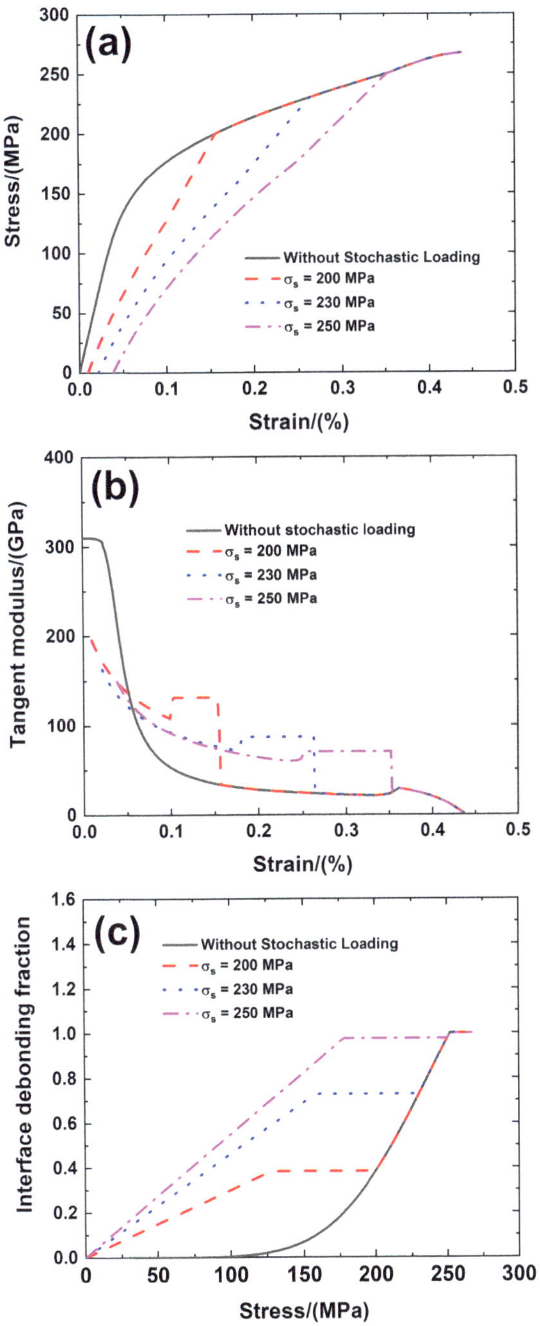

Fig. 1.5 **a** Tensile stress–strain curves; **b** the composite tangent modulus versus applied strain curves; **c** the interface debonding ratio versus applied stress curves; and **d** the broken fibers fraction versus applied stress curves of SiC/SiC composite under $\sigma_s = 200$, 230, and 250 MPa when $\tau_i = 15$ MPa

Fig. 1.5 (continued)

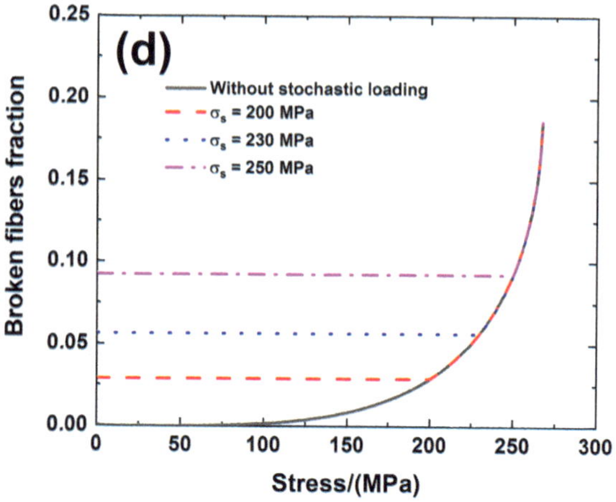

$= 0.727$. When the applied stress increases from $\sigma_{\mathrm{tr}} = 154$ MPa to $\sigma = 230$ MPa, the composite tangent modulus remains constant value of $E_{\mathrm{p}} = 86.8$ GPa with the interface debonding ratio of $\eta = 0.72$.

- Under $\sigma_{\mathrm{s}} = 250$ MPa, the initial composite strain is $\varepsilon_0 = 0.038\%$ due to the damage modes of the matrix cracking and interface debonding; the composite initial tangent modulus is approximately $E_{\mathrm{p}} = 149.5$ GPa, the degradation rate of the composite tangent modulus is 52% compared with original specimen, and the fibers broken fraction is $q = 0.092$. With increasing applied stress to $\sigma_{\mathrm{tr}} = 173.8$ MPa, the interface debonding ratio increases, and the composite tangent modulus decreases to approximately $E_{\mathrm{p}} = 60.7$ GPa corresponding to the interface debonding ratio of $\eta = 0.952$. When the applied stress increases from $\sigma_{\mathrm{tr}} = 173.8$ MPa to $\sigma = 250$ MPa, the composite tangent modulus remains constant value of approximately $E_{\mathrm{p}} = 70.4$ GPa with the interface debonding ratio of $\eta = 0.975$.

When $\tau_{\mathrm{i}} = 20$ MPa, the composite damage mechanisms and tensile damage evolution and fracture subjected to different stochastic loading stresses of $\sigma_{\mathrm{s}} = 200$, 230, and 250 MPa are analyzed, as following:

- Under $\sigma_{\mathrm{s}} = 200$ MPa, the damage modes of matrix cracking and interface debonding occur, leading to the increase of the composite initial strain, decreasing of the composite tangent modulus, and increase of the broken fibers fraction. The initial composite strain is $\varepsilon_0 = 0.00943\%$; the composite initial tangent modulus is approximately $E_{\mathrm{p}} = 215$ GPa, the degradation rate of composite tangent modulus is 31% compared with original specimen, and the fibers broken fraction is $q = 0.028$. With increasing applied stress to $\sigma_{\mathrm{tr}} = 123.2$ MPa, the interface debonding ratio increases, and the composite tangent modulus decreases to approximately $E_{\mathrm{p}} = 130$ GPa corresponding to the interface debonding ratio of $\eta = 0.275$. When the applied stress increases from $\sigma_{\mathrm{tr}} = 123.2$ MPa to $\sigma = 200$ MPa, the composite

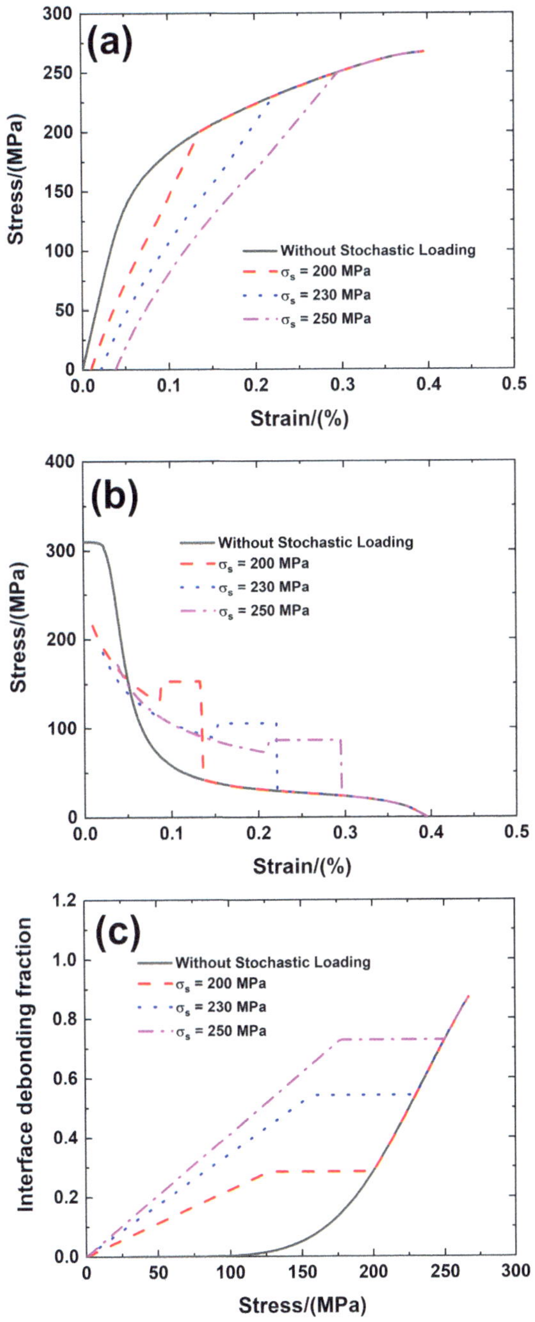

Fig. 1.6 **a** Tensile stress–strain curves; **b** the composite tangent modulus versus applied strain curves; **c** the interface debonding ratio versus applied stress curves; and **d** the broken fibers fraction versus applied stress curves of SiC/SiC composite under $\sigma_s = 200$, 230, and 250 MPa when $\tau_i = 20$ MPa

Fig. 1.6 (continued)

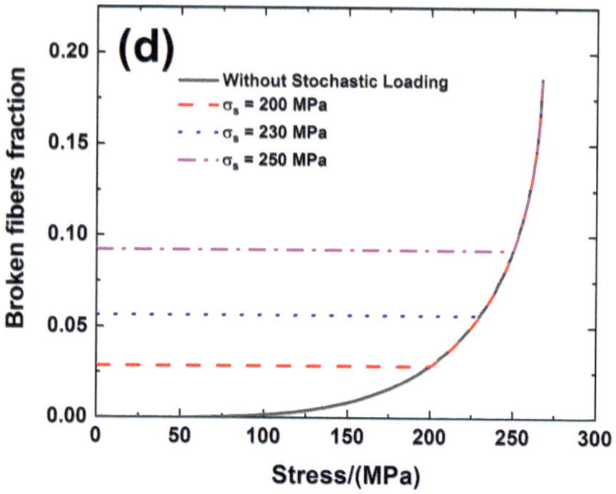

Table 1.2 Effect of interface shear stress ($\tau_i = 15$, and 20 MPa) on tensile damage and fracture of SiC/SiC composite subjected to $\sigma_s = 200$, 230, and 250 MPa

$\tau_i = 15$ MPa	$\sigma_s = 200$ MPa			
	$\varepsilon_0/(\%)$	$E_p/(\text{GPa})$	$\sigma_{tr}/(\text{MPa})$	$q/(\%)$
	0.00943	195.6	125.4	0.028
	$\sigma_s = 230$ MPa			
	$\varepsilon_0/(\%)$	$E_p/(\text{GPa})$	$\sigma_{tr}/(\text{MPa})$	$q/(\%)$
	0.021	163	154	0.056
	$\sigma_s = 250$ MPa			
	$\varepsilon_0/(\%)$	$E_p/(\text{GPa})$	$\sigma_{tr}/(\text{MPa})$	$q/(\%)$
	0.038	149.5	173.8	0.092
$\tau_i = 20$ MPa	$\sigma_s = 200$ MPa			
	$\varepsilon_0/(\%)$	$E_p/(\text{GPa})$	$\sigma_{tr}/(\text{MPa})$	$q/(\%)$
	0.00943	215	125.4	0.028
	$\sigma_s = 230$ MPa			
	$\varepsilon_0/(\%)$	$E_p/(\text{GPa})$	$\sigma_{tr}/(\text{MPa})$	$q/(\%)$
	0.021	185	154	0.056
	$\sigma_s = 250$ MPa			
	$\varepsilon_0/(\%)$	$E_p/(\text{GPa})$	$\sigma_{tr}/(\text{MPa})$	$q/(\%)$
	0.038	171.7	173.8	0.092

tangent modulus remains constant value of approximately $E_p = 152.9$ GPa with the interface debonding ratio of $\eta = 0.284$.

- Under $\sigma_s = 230$ MPa, the initial composite strain is $\varepsilon_0 = 0.021\%$ due to the damage modes of the matrix cracking and interface debonding; the initial composite tangent modulus is $E_p = 185$ GPa, the degradation rate of the composite tangent modulus is 40% compared with original specimen, and the fibers broken fraction is $q = 0.056$. With increasing applied stress to $\sigma_{tr} = 154$ MPa, the interface debonding ratio increases, and the composite tangent modulus decreases to approximately $E_p = 88.6$ GPa corresponding to the interface debonding ratio of $\eta = 0.53$. When the applied stress increases from $\sigma_{tr} = 154$ MPa to $\sigma = 230$ MPa, the composite tangent modulus remains constant value of approximately $E_p = 105.5$ GPa with the interface debonding ratio of $\eta = 0.54$.

- Under $\sigma_s = 250$ MPa, the initial composite strain is $\varepsilon_0 = 0.038\%$ due to the damage modes of the matrix cracking and interface debonding; the composite initial tangent modulus is approximately $E_p = 171.7$ GPa, the degradation rate of composite tangent modulus is 45% compared with original specimen, and the fibers broken fraction is $q = 0.092$. With increasing applied stress to $\sigma_{tr} = 173.8$ MPa, the interface debonding ratio increases, and the composite tangent modulus decreases to approximately $E_p = 72.8$ GPa corresponding to the interface debonding ratio of $\eta = 0.714$. When the applied stress increases from $\sigma_{tr} = 173.8$ MPa to $\sigma = 250$ MPa, the composite tangent modulus remains constant value of approximately $E_p = 86.3$ GPa with the interface debonding ratio of $\eta = 0.73$.

1.3.3 Tensile Damage and Fracture of SiC/SiC Composite with Stochastic Loading for Different Interface Debonding Energy

The interface debonding energy is a key interface property of fiber-reinforced CMCs. Domergue et al. (1995) estimated the interface debonding energy of unidirectional SiC/CAS composite by analyzing the hysteresis loops and obtained the interface debonding energy which lies in the range of $\Gamma_d = 0.1$–0.8 J/m^2. Effect of the interface debonding energy (i.e., $\Gamma_d = 0.1$ and 0.3 J/m^2) on the tensile damage and fracture of SiC/SiC composite subjected to $\sigma_s = 180$, 220, and 250 MPa are shown in Figs. 1.7 and 1.8 and Table 1.3. When the interface debonding energy increases, the initial composite strain, composite tangent modulus, and broken fibers fraction remain the same, and the transition stress for interface debonding decreases.

When $\Gamma_d = 0.1$ J/m^2, the composite damage mechanisms and tensile damage evolution and fracture subjected to different stochastic loading stresses of $\sigma_s = 180$, 220, and 250 MPa are analyzed, as following:

- Under $\sigma_s = 180$ MPa, the damage modes of matrix cracking and interface debonding occur, leading to the increase of the composite initial strain, decreasing of the composite tangent modulus, and increase of the broken fibers fraction. The initial composite strain is $\varepsilon_0 = 0.00537\%$; the composite initial tangent modulus is approximately $E_p = 251.5$ GPa, the degradation rate of the composite tangent

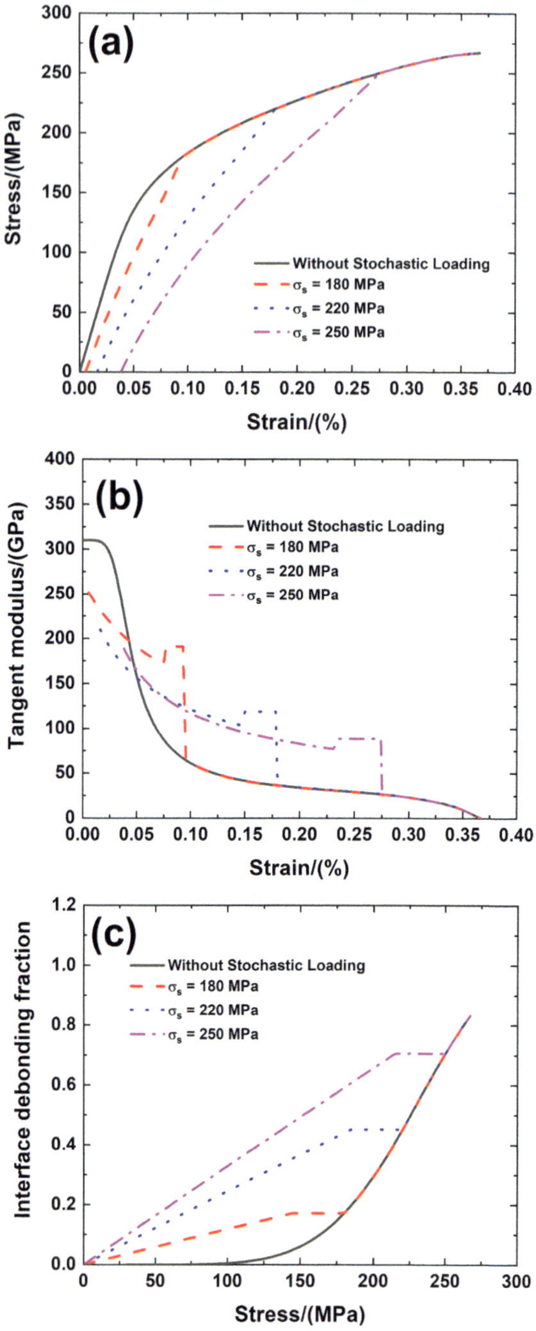

Fig. 1.7 **a** Tensile stress–strain curves; **b** the composite tangent modulus versus applied strain curves; **c** the interface debonding ratio versus applied stress curves; and **d** the broken fibers fraction versus applied stress curves of SiC/SiC composite under $\sigma_s = 180$, 220, and 250 MPa when $\Gamma_d = 0.1$ J/m^2

Fig. 1.7 (continued)

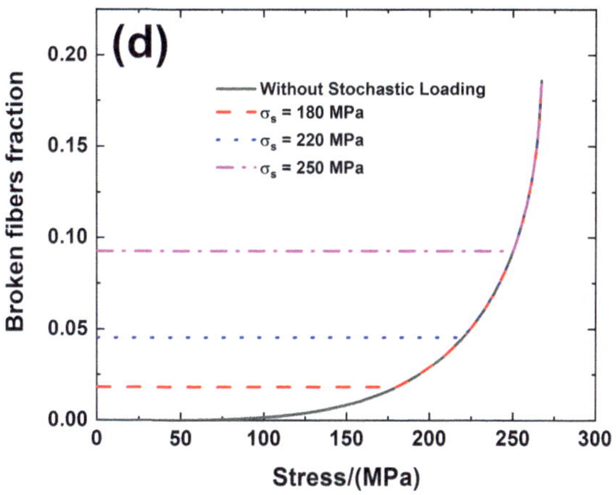

modulus is 19% compared with original specimen, and the fibers broken fraction is $q = 0.018$. With increasing applied stress to $\sigma_{\mathrm{tr}} = 140.8$ MPa, the interface debonding ratio increases, and the composite tangent modulus decreases to approximately $E_{\mathrm{p}} = 171.6$ GPa corresponding to the interface debonding ratio of $\eta = 0.167$. When the applied stress increases from $\sigma_{\mathrm{tr}} = 140.8$ MPa to $\sigma = 180$ MPa, the composite tangent modulus remains constant value of approximately $E_{\mathrm{p}} = 191$ GPa with the interface debonding ratio of $\eta = 0.171$.

- Under $\sigma_{\mathrm{s}} = 220$ MPa, the initial composite strain is $\varepsilon_0 = 0.0161\%$ due to the damage modes of the matrix cracking and interface debonding; the initial composite tangent modulus is approximately $E_{\mathrm{p}} = 209$ GPa, the degradation rate of the composite tangent modulus is 33% compared with original specimen, and the fibers broken fraction is $q = 0.045$. With increasing applied stress to $\sigma_{\mathrm{tr}} = 182.6$ MPa, the interface debonding ratio increases, and the composite tangent modulus decreases to approximately $E_{\mathrm{p}} = 103.6$ GPa corresponding to the interface debonding ratio of $\eta = 0.446$. When the applied stress increases from $\sigma_{\mathrm{tr}} = 182.6$ MPa to $\sigma = 220$ MPa, the composite tangent modulus remains constant value of approximately $E_{\mathrm{p}} = 118.4$ GPa with the interface debonding ratio of $\eta = 0.451$.
- Under $\sigma_{\mathrm{s}} = 250$ MPa, the initial composite strain is $\varepsilon_0 = 0.038\%$ due to the damage modes of the matrix cracking and interface debonding; the composite initial tangent modulus is approximately $E_{\mathrm{p}} = 188.5$ GPa, the degradation rate of the composite tangent modulus is 40% compared with the original specimen, and the fibers broken fraction is $q = 0.092$. With increasing applied stress to $\sigma_{\mathrm{tr}} = 211.2$ MPa, the interface debonding ratio increases, and the composite tangent modulus decreases to approximately $E_{\mathrm{p}} = 77$ GPa corresponding to the interface debonding ratio of $\eta = 0.694$. When the applied stress increases from $\sigma_{\mathrm{tr}} = 211.2$ MPa to $\sigma = 250$ MPa, the composite tangent modulus remains constant value of approximately $E_{\mathrm{p}} = 88.2$ GPa with the interface debonding ratio of $\eta = 0.705$.

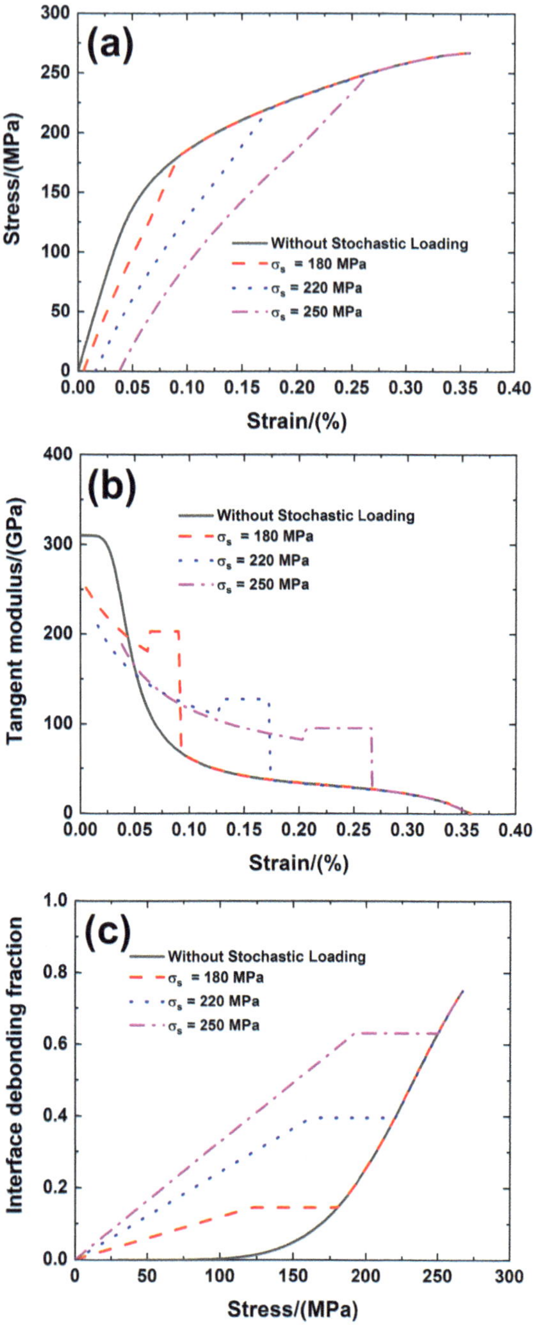

Fig. 1.8 **a** Tensile stress–strain curves; **b** the composite tangent modulus versus applied strain curves; **c** the interface debonding ratio versus applied stress curves; and **d** the broken fibers fraction versus applied stress curves of SiC/SiC composite under $\sigma_s = 180$, 220, and 250 MPa when $\Gamma_d = 0.3$ J/m^2

Fig. 1.8 (continued)

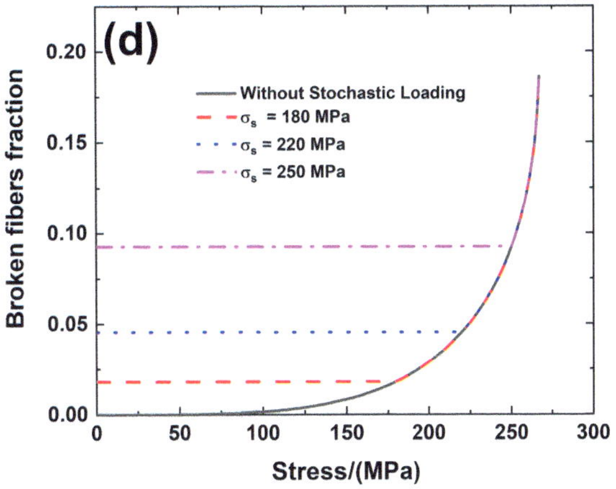

Table 1.3 Effect of interface debonding energy (i.e., $\Gamma_d = 0.1$ and $0.3\ \text{J/m}^2$) on tensile damage and fracture of SiC/SiC composite under $\sigma_s = 180$, 220, and 250 MPa

$\Gamma_d = 0.1\ \text{J/m}^2$	$\sigma_s = 180$ MPa			
	ε_0/(%)	E_p/(GPa)	σ_{tr}/(MPa)	q/(%)
	0.00537	251.5	140.8	0.018
	$\sigma_s = 220$ MPa			
	ε_0/(%)	E_p/(GPa)	σ_{tr}/(MPa)	q/(%)
	0.0161	209	182.6	0.045
	$\sigma_s = 250$ MPa			
	ε_0/(%)	E_p/(GPa)	σ_{tr}/(MPa)	q/(%)
	0.038	188.5	211.2	0.092
$\Gamma_d = 0.3\ \text{J/m}^2$	$\sigma_s = 180$ MPa			
	ε_0/(%)	E_p/(GPa)	σ_{tr}/(MPa)	q/(%)
	0.00537	251.5	118.8	0.018
	$\sigma_s = 220$ MPa			
	ε_0/(%)	E_p/(GPa)	σ_{tr}/(MPa)	q/(%)
	0.0161	209	158.4	0.045
	$\sigma_s = 250$ MPa			
	ε_0/(%)	E_p/(GPa)	σ_{tr}/(MPa)	q/(%)
	0.038	188.5	189.2	0.092

When $\Gamma_d = 0.3\ \text{J/m}^2$, the composite damage mechanisms and tensile damage evolution and fracture subjected to different stochastic loading stresses of $\sigma_s = 180$, 220, and 250 MPa are analyzed, as following:

- Under $\sigma_s = 180$ MPa, the damage modes of matrix cracking and interface debonding occur, leading to the increase of the composite initial strain, decreasing of the composite tangent modulus, and increase of the broken fibers fraction. The initial composite strain is $\varepsilon_0 = 0.00537\%$; the initial composite tangent modulus is approximately $E_p = 251.5$ GPa, the degradation rate of the composite tangent modulus is 19% compared with the original specimen, and the fibers broken fraction is $q = 0.018$. With increasing applied stress to $\sigma_{tr} = 118.8$ MPa, the interface debonding ratio increases, and the composite tangent modulus decreases to approximately $E_p = 180.6$ GPa corresponding to the interface debonding ratio of $\eta = 0.141$. When the applied stress increases from $\sigma_{tr} = 118.8$ MPa to $\sigma =$ 180 MPa, the composite tangent modulus remains constant value of approximately $E_p = 202$ GPa with the interface debonding ratio of $\eta = 0.144$.

- Under $\sigma_s = 220$ MPa, the initial composite strain is $\varepsilon_0 = 0.0161\%$; the initial composite tangent modulus is approximately $E_p = 209$ GPa, the degradation rate of composite tangent modulus is 33% compared with original specimen, and the fibers broken fraction is $q = 0.045$. With increasing applied stress to $\sigma_{tr} = 158.4$ MPa, the interface debonding ratio increases, and the composite tangent modulus decreases to approximately $E_p = 110.4$ GPa corresponding to the interface debonding ratio of $\eta = 0.387$. When the applied stress increases from $\sigma_{tr} = 158.4$ MPa to $\sigma = 220$ MPa, the composite tangent modulus remains constant value of approximately $E_p = 127.6$ GPa with the interface debonding ratio of $\eta = 0.396$.
- Under $\sigma_s = 250$ MPa, the initial composite strain is $\varepsilon_0 = 0.038\%$ due to the damage modes of the matrix cracking and interface debonding; the composite initial tangent modulus is approximately $E_p = 188.5$ GPa, the degradation rate of composite tangent modulus is 40% compared with the original specimen, and the fibers broken fraction is $q = 0.092$. With increasing applied stress to $\sigma_{tr} =$ 189.2 MPa, the interface debonding ratio increases, and the composite tangent modulus decreases to approximately $E_p = 82$ GPa corresponding to the interface debonding ratio of $\eta = 0.622$. When the applied stress increases from $\sigma_{tr} =$ 189.2 MPa to $\sigma = 250$ MPa, the composite tangent modulus remains constant value of approximately $E_p = 95.1$ GPa with the interface debonding ratio of $\eta =$ 0.63.

1.3.4 Tensile Damage and Fracture of SiC/SiC Composite with Stochastic Loading for Different Saturation Matrix Crack Spacing

Li (2020b) investigated multiple matrix cracking of CMCs with different fiber preforms and found that the saturation matrix cracking spacing is in the range of $l_s = 100$ and 500 μm. In the present analysis, effect of the saturation matrix crack spacing (i.e., $l_s = 200$ and 250 μm) on the tensile damage and fracture of SiC/SiC composite subjected to $\sigma_s = 180$, 220, and 250 MPa are shown in Figs. 1.9 and 1.10 and Table 1.4. When saturation matrix crack spacing increases, the initial composite

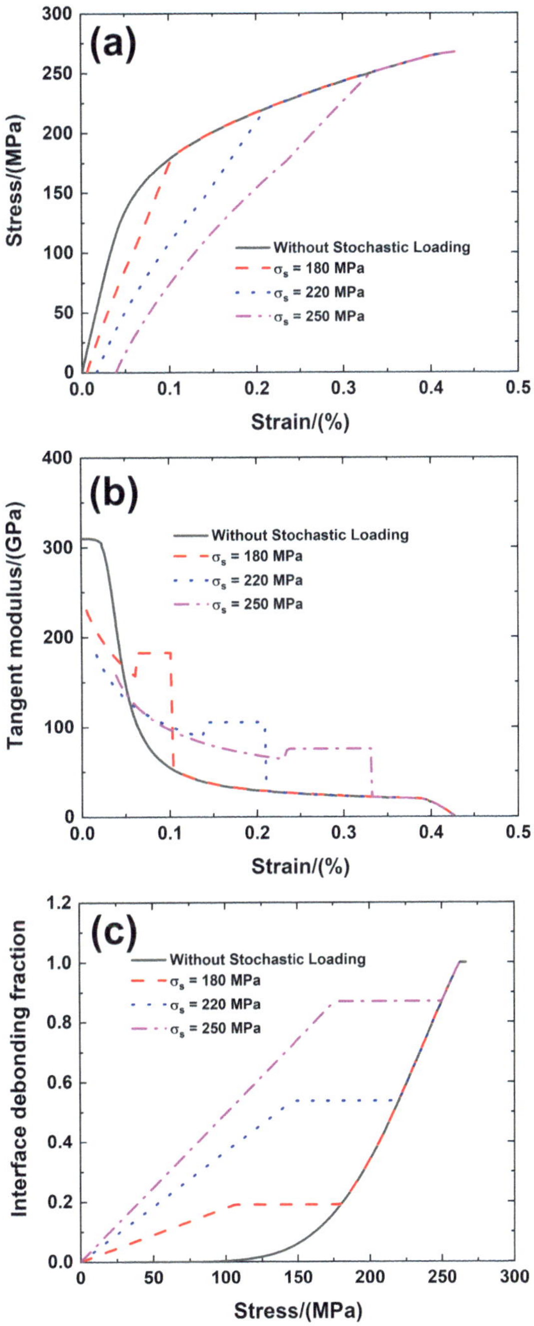

Fig. 1.9 **a** Tensile stress–strain curves; **b** the composite tangent modulus versus applied strain curves; **c** the interface debonding ratio versus applied stress curves; and **d** the broken fibers fraction versus applied stress curves of SiC/SiC composite under $\sigma_s = 180$, 220, and 250 MPa when $l_s = 200\ \mu m$

Fig. 1.9 (continued)

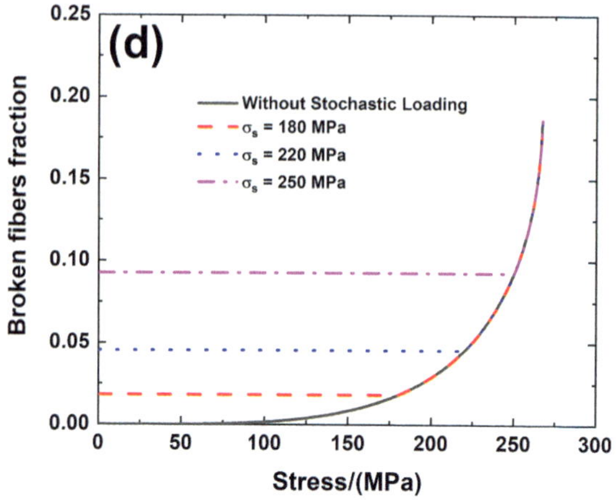

strain decreases, the initial composite tangent modulus increases, the transition stress for interface debonding and initial fibers broken fraction remain the same.

When $l_s = 200$ μm, the composite damage mechanisms and tensile damage evolution and fracture subjected to different stochastic loading stresses of $\sigma_s = 180$, 220, and 250 MPa are analyzed, as following:

- Under $\sigma_s = 180$ MPa, the damage modes of matrix cracking and interface debonding occur, leading to the increase of the composite initial strain, decreasing of the composite tangent modulus, and increase of the broken fibers fraction. The initial composite strain is $\varepsilon_0 = 0.00567\%$ due to the damage modes of the matrix cracking and interface debonding; the composite initial tangent modulus is approximately $E_p = 229.8$ GPa, the degradation rate of composite tangent modulus is 26% compared with the original specimen, and the fibers broken fraction is $q = 0.018$. With increasing applied stress to $\sigma_{tr} = 103.4$ MPa, the interface debonding ratio increases, and the composite tangent modulus decreases to approximately $E_p = 156.5$ GPa corresponding to the interface debonding ratio of $\eta = 0.184$. When the applied stress increases from $\sigma_{tr} = 103.4$ MPa to $\sigma =$ 180 MPa, the composite tangent modulus remains constant value of approximately $E_p = 182.3$ GPa with the interface debonding ratio of $\eta = 0.189$.
- Under $\sigma_s = 220$ MPa, the initial composite strain is $\varepsilon_0 = 0.0168\%$ due to the damage modes of the matrix cracking and the interface debonding; the initial composite tangent modulus is approximately $E_p = 180.3$ GPa, the degradation rate of composite tangent modulus is 42% compared with original specimen, and the fibers broken fraction is $q = 0.045$. With increasing applied stress to σ_{tr} = 143 MPa, the interface debonding ratio increases, and the composite tangent modulus decreases to approximately $E_p = 88.1$ GPa corresponding to the interface debonding ratio of $\eta = 0.524$. When the applied stress increases from $\sigma_{tr} =$ 143 MPa to $\sigma = 220$ MPa, the composite tangent modulus remains constant

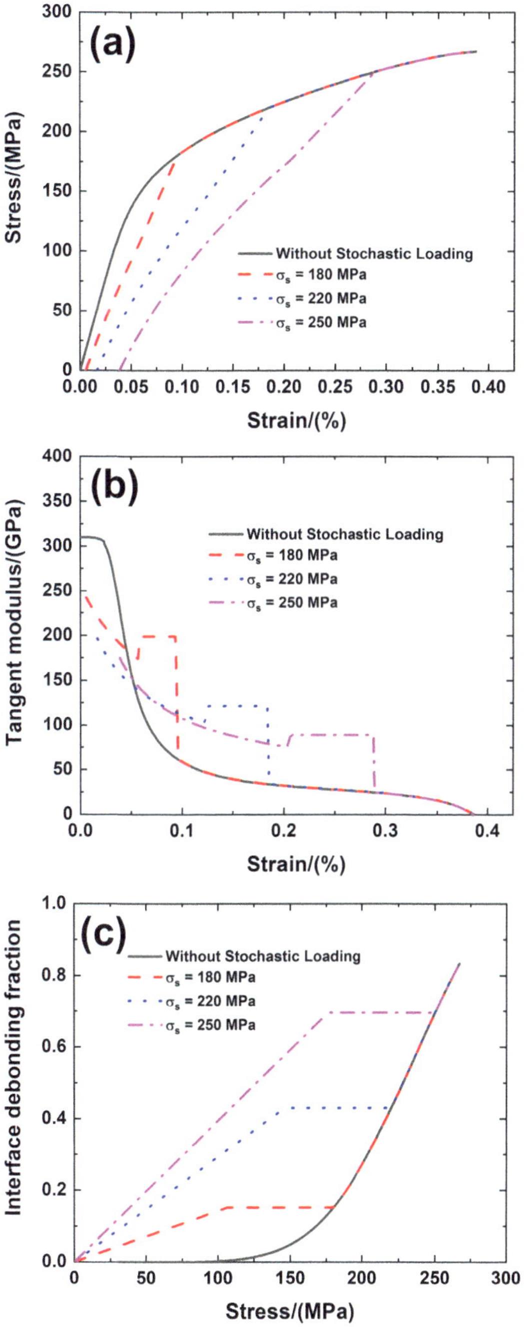

Fig. 1.10 **a** Tensile stress–strain curves; **b** the composite tangent modulus versus applied strain curves; **c** the interface debonding ratio versus applied stress curves; and **d** the broken fibers fraction versus applied stress curves of SiC/SiC composite under $\sigma_s = 180$, 220, and 250 MPa when $l_s = 250\ \mu m$

Fig. 1.10 (continued)

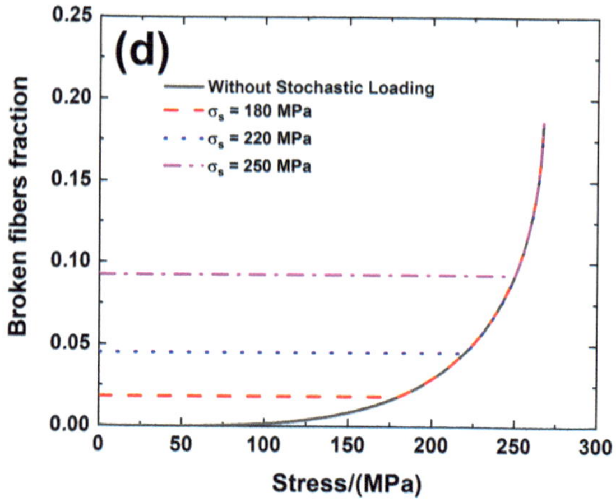

Table 1.4 Effect of saturation matrix crack spacing (i.e., $l_s = 200$ and 250 μm) on tensile damage and fracture of SiC/SiC composite subjected to $\sigma_s = 180$, 220, and 250 MPa

$l_s = 200$ μm	$\sigma_s = 180$ MPa			
	$\varepsilon_0/(\%)$	$E_p/(\text{GPa})$	$\sigma_{tr}/(\text{MPa})$	$q/(\%)$
	0.00567	229.8	103.4	0.018
	$\sigma_s = 220$ MPa			
	$\varepsilon_0/(\%)$	$E_p/(\text{GPa})$	$\sigma_{tr}/(\text{MPa})$	$q/(\%)$
	0.0168	180.3	143	0.045
	$\sigma_s = 250$ MPa			
	$\varepsilon_0/(\%)$	$E_p/(\text{GPa})$	$\sigma_{tr}/(\text{MPa})$	$q/(\%)$
	0.038	157.7	173.8	0.092
$l_s = 250$ μm	$\sigma_s = 180$ MPa			
	$\varepsilon_0/(\%)$	$E_p/(\text{GPa})$	$\sigma_{tr}/(\text{MPa})$	$q/(\%)$
	0.00549	242.3	103.4	0.018
	$\sigma_s = 220$ MPa			
	$\varepsilon_0/(\%)$	$E_p/(\text{GPa})$	$\sigma_{tr}/(\text{MPa})$	$q/(\%)$
	0.0164	196.8	143	0.045
	$\sigma_s = 250$ MPa			
	$\varepsilon_0/(\%)$	$E_p/(\text{GPa})$	$\sigma_{tr}/(\text{MPa})$	$q/(\%)$
	0.038	174.8	173.8	0.092

value of approximately $E_p = 105.2$ GPa with the interface debonding ratio of $\eta = 0.536$.

- Under $\sigma_s = 250$ MPa, the initial composite strain is $\varepsilon_0 = 0.038\%$ due to the damage modes of the matrix cracking and the interface debonding; the initial composite tangent modulus is approximately $E_p = 157.7$ GPa, the degradation rate of composite tangent modulus is 50% compared with original specimen, and the fibers broken fraction is $q = 0.092$. With increasing applied stress to $\sigma_{\text{tr}} = 173.8$ MPa, the interface debonding ratio increases and the composite tangent modulus decreases to approximately $E_{\text{p}} = 64$ GPa corresponding to thee interface debonding ratio of $\eta = 0.857$. When the applied stress increases from $\sigma_{\text{tr}} = 173.8$ MPa to $\sigma = 250$ MPa, the composite tangent modulus remains constant value of approximately $E_{\text{p}} = 75.3$ GPa with $\eta = 0.869$.

When $l_{\text{s}} = 250\ \mu\text{m}$, the composite damage mechanisms and tensile damage evolution and fracture subjected to different stochastic loading stresses of $\sigma_{\text{s}} = 180$, 220, and 250 MPa are analyzed, as following:

- Under $\sigma_{\text{s}} = 180$ MPa, the damage modes of matrix cracking and interface debonding occur, leading to the increase of the composite initial strain, decreasing of the composite tangent modulus, and increase of the broken fibers fraction. The initial composite strain is $\varepsilon_0 = 0.00549\%$ due to the damages of the matrix cracking and the interface debonding; the initial composite tangent modulus is approximately $E_{\text{p}} = 242.3$ GPa, the degradation rate of composite tangent modulus is 22% compared with original specimen, and the fibers broken fraction is $q = 0.018$. With increasing applied stress to $\sigma_{\text{tr}} = 103.4$ MPa, the interface debonding ratio increases, and the composite tangent modulus decreases to approximately $E_{\text{p}} = 173.7$ GPa corresponding to the interface debonding ratio of $\eta = 0.147$. When the applied stress increases from $\sigma_{\text{tr}} = 103.4.5$ MPa to $\sigma = 180$ MPa, the composite tangent modulus remains constant value of approximately $E_{\text{p}} = 198.7$ GPa with the interface debonding ratio of $\eta = 0.151$.
- Under $\sigma_{\text{s}} = 220$ MPa, the initial composite strain is $\varepsilon_0 = 0.0164\%$ due to the damage modes of the matrix cracking and the interface debonding; the initial composite tangent modulus is approximately $E_{\text{p}} = 196.8$ GPa, the degradation rate of composite tangent modulus is 37% compared with original specimen, and the fibers broken fraction is $q = 0.045$. With increasing applied stress to $\sigma_{\text{tr}} = 143$ MPa, the interface debonding ratio increases and the composite tangent modulus decreases to approximately $E_{\text{p}} = 102.8$ GPa corresponding to the interface debonding ratio of $\eta = 0.42$. When the applied stress increases from $\sigma_{\text{tr}} = 143$ MPa to $\sigma = 220$ MPa, the composite tangent modulus remains constant value of approximately $E_{\text{p}} = 121.2$ GPa with the interface debonding ratio of $\eta = 0.43$.
- Under $\sigma_{\text{s}} = 250$ MPa, the initial composite strain is $\varepsilon_0 = 0.038\%$ due to the damage modes of the matrix cracking and the interface debonding; the initial composite tangent modulus is approximately $E_{\text{p}} = 174.8$ GPa, the degradation rate of composite tangent modulus is 44% compared with original specimen, and the fibers broken fraction is $q = 0.092$. With increasing applied stress to $\sigma_{\text{tr}} = 173.8$ MPa, the interface debonding ratio increases, and the composite tangent modulus decreases to approximately $E_{\text{p}} = 75$ GPa corresponding to the interface debonding ratio of $\eta = 0.685$. When the applied stress increases from $\sigma_{\text{tr}} = 173.8$ MPa to $\sigma = 250$ MPa, the composite tangent modulus remains constant value of approximately $E_{\text{p}} = 88.6$ GPa with the interface debonding ratio of $\eta = 0.695$.

1.3.5 Tensile Damage and Fracture of SiC/SiC Composite with Stochastic Loading for Different Fiber Strength

Guo et al. (1998) investigated the SiC fiber strength and found that the SiC fiber strength is in the range between $\sigma_{fc} = 2.3$ and 3.7 GPa. In the present analysis, effect of the fibers strength (i.e., $\sigma_{fc} = 2.0$ and 2.5 GPa) on the tensile damage and fracture of SiC/SiC composite subjected to $\sigma_s = 180$, 220, and 250 MPa are shown in Figs. 1.11 and 1.12 and Table 1.5. When the fiber strength increases, the initial composite strain and fiber broken fraction decrease, and the initial tangent composite modulus and transition stress for interface debonding remain the same.

When $\sigma_{fc} = 2.0$ GPa, the composite damage mechanisms and tensile damage evolution and fracture subjected to different stochastic loading stresses of $\sigma_s = 180$, 220, and 250 MPa are analyzed, as following:

- Under $\sigma_s = 180$ MPa, the damage modes of matrix cracking and interface debonding occur, leading to the increase of the composite initial strain, decreasing of the composite tangent modulus, and increase of the broken fibers fraction. The composite initial strain is $\varepsilon_0 = 0.0065\%$ due to the damages of the matrix cracking and the interface debonding; the initial composite tangent modulus is approximately $E_p = 251.5$ GPa, the degradation rate of composite tangent modulus is 19% compared with original specimen, and the fibers broken fraction is $q = 0.022$. With increasing applied stress to $\sigma_{tr} = 103.4$ MPa, the interface debonding ratio increases, and the composite tangent modulus decreases to approximately $E_p =$ 187.4 GPa corresponding to the interface debonding ratio of $\eta = 0.122$. When the applied stress increases from $\sigma_{tr} = 103.4$ MPa to $\sigma = 180$ MPa, the composite tangent modulus remains constant value of approximately $E_p = 211.3$ GPa with $\eta = 0.126$.
- Under $\sigma_s = 220$ MPa, the initial composite strain is $\varepsilon_0 = 0.02\%$ due to the damage modes of the matrix cracking and the interface debonding; the initial composite tangent modulus is approximately $E_p = 209.6$ GPa, the degradation rate of composite tangent modulus is 33% compared with original specimen, and the fibers broken fraction is $q = 0.058$. With increasing stress to $\sigma_{tr} = 143$ MPa, the interface debonding ratio increases, and the composite tangent modulus decreases to approximately $E_p = 115.7$ GPa corresponding to the interface debonding ratio of $\eta = 0.349$. When the applied stress increases from $\sigma_{tr} = 143$ MPa to $\sigma =$ 220 MPa, the composite tangent modulus remains constant value of approximately $E_p = 134.9$ GPa with the interface debonding ratio of $\eta = 0.357$.
- Under $\sigma_s = 250$ MPa, the initial composite strain is $\varepsilon_0 = 0.059\%$; the composite initial tangent modulus is approximately $E_p = 188.5$ GPa, the degradation rate of composite tangent modulus is 40% compared with original specimen, and the fibers broken fraction is $q = 0.138$. With increasing applied stress to $\sigma_{tr} =$ 173.8 MPa, the interface debonding ratio increases and the composite tangent modulus decreases to approximately $E_p = 86$ GPa corresponding to the interface debonding ratio of $\eta = 0.571$. When the applied stress increases from $\sigma_{tr} =$ 173.8 MPa to $\sigma = 250$ MPa, the composite tangent modulus remains constant

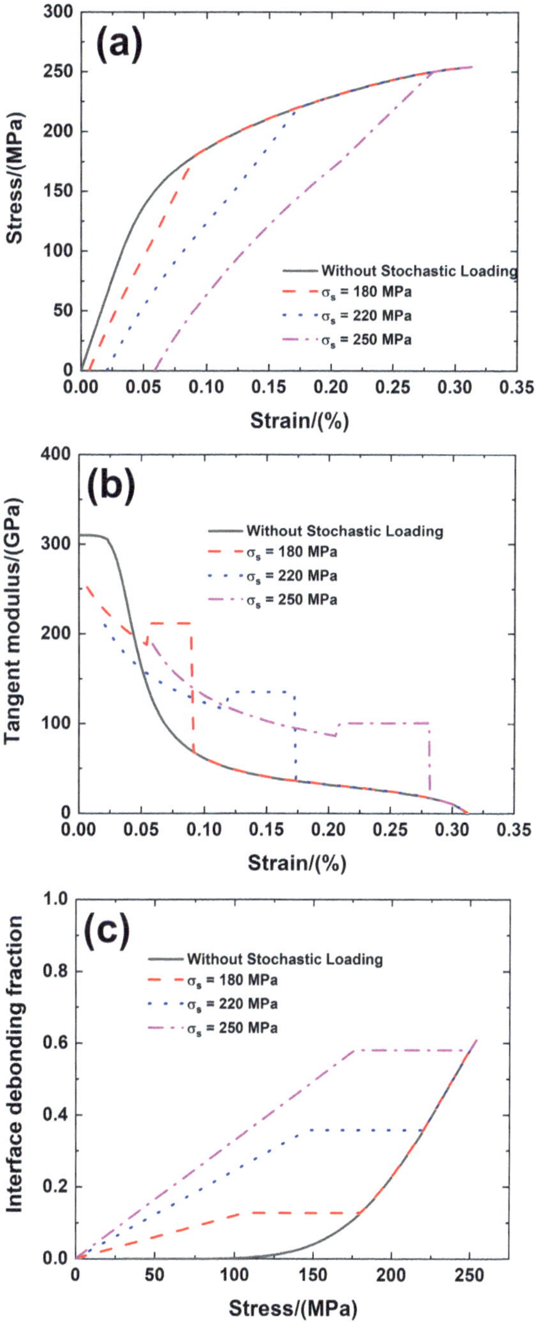

Fig. 1.11 **a** Tensile stress–strain curves; **b** the composite tangent modulus versus applied strain curves; **c** the interface debonding ratio versus applied stress curves; and **d** the broken fibers fraction versus applied stress curves of SiC/SiC composite under $\sigma_s = 180$, 220, and 250 MPa when $\sigma_{fc} = 2.0$ GPa

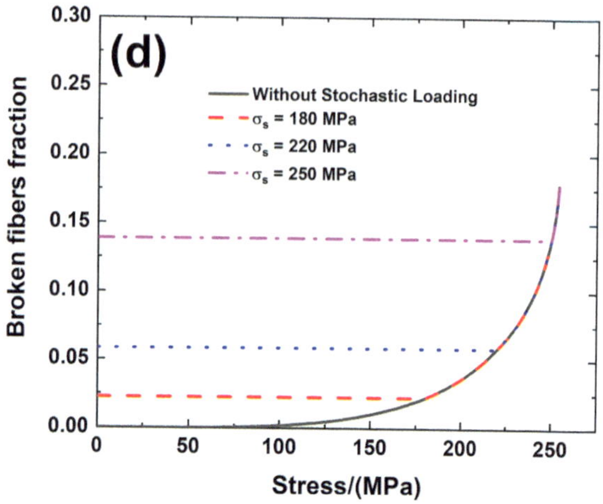

Fig. 1.11 (continued)

value of approximately $E_p = 100.6$ GPa with the interface debonding ratio of $\eta = 0.579$.

When $\sigma_{fc} = 2.5$ GPa, the composite damage mechanisms and tensile damage evolution and fracture subjected to different stochastic loading stresses of $\sigma_s = 180$, 220, and 250 MPa are analyzed, as following:

- Under $\sigma_s = 180$ MPa, the damage modes of matrix cracking and interface debonding occur, leading to the increase of the composite initial strain, decreasing of the composite tangent modulus, and increase of the broken fibers fraction. The initial composite strain is $\varepsilon_0 = 0.0028\%$ due to the damages of the matrix cracking and the interface debonding; the composite initial tangent modulus is approximately $E_p = 251.5$ GPa, the degradation rate of composite tangent modulus is 19% compared with original specimen, and the fibers broken fraction is $q = 0.0087$. With increasing applied stress to $\sigma_{tr} = 103.4$ MPa, the interface debonding ratio increases, and the composite tangent modulus decreases to approximately $E_p = 187.4$ GPa corresponding to the interface debonding ratio of $\eta = 0.122$. When the applied stress increases from $\sigma_{tr} = 103.4$ MPa to $\sigma = 180$ MPa, the composite tangent modulus remains constant value of approximately $E_p = 211.3$ GPa with $\eta = 0.126$.
- Under $\sigma_s = 220$ MPa, the initial composite strain is $\varepsilon_0 = 0.007\%$ due to the damages of the matrix cracking and the interface debonding; the initial composite tangent modulus is approximately $E_p = 209.6$ GPa, the degradation rate of composite tangent modulus is 33% compared with original specimen, and the fibers broken fraction is $q = 0.02$. With increasing applied stress to $\sigma_{tr} = 143$ MPa, the interface debonding ratio increases, and the composite tangent modulus decreases to approximately $E_p = 115.7$ GPa corresponding to $\eta = 0.344$. When the applied stress increases from $\sigma_{tr} = 140.8$ MPa to $\sigma = 220$ MPa, the composite tangent modulus remains constant value of approximately $E_p = 134.9$ GPa with $\eta = 0.357$.

Fig. 1.12 **a** Tensile stress–strain curves; **b** the composite tangent modulus versus applied strain curves; **c** the interface debonding ratio versus applied stress curves; and **d** the broken fibers fraction versus applied stress curves of SiC/SiC composite under $\sigma_s = 180$, 220, and 250 MPa when $\sigma_{fc} = 2.5$ GPa

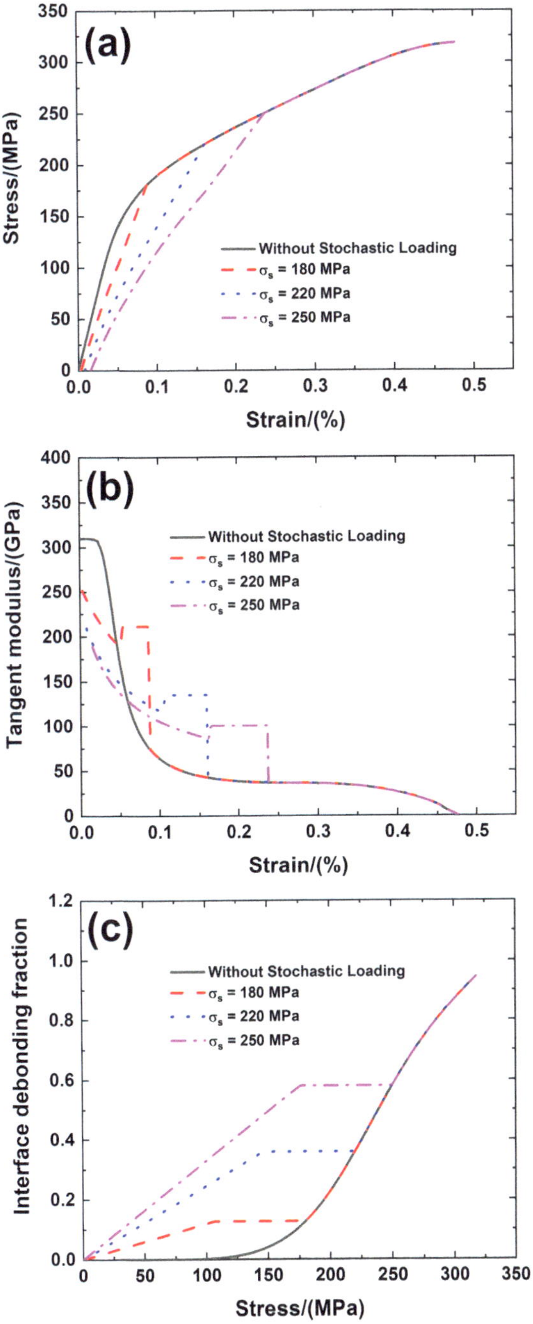

Fig. 1.12 (continued)

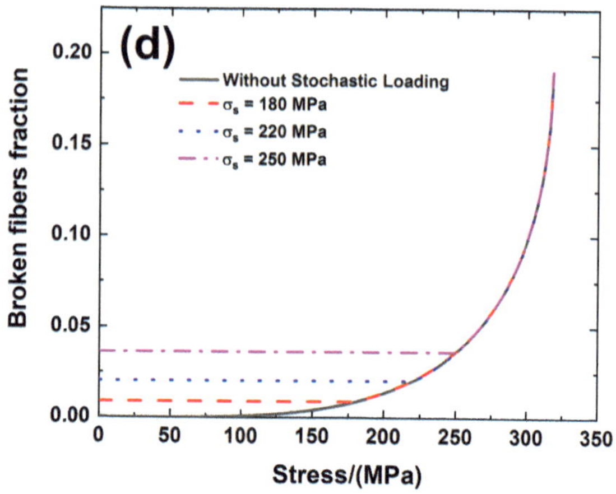

Table 1.5 Effect of fiber strength (i.e., $\sigma_{fc} = 2.0$, and 2.5 GPa) on tensile damage and fracture of SiC/SiC composite under $\sigma_s = 180$, 220, and 250 MPa

$\sigma_{fc} = 2.0$ GPa	$\sigma_s = 180$ MPa			
	$\varepsilon_0/(\%)$	$E_p/(\text{GPa})$	$\sigma_{tr}/(\text{MPa})$	$q/(\%)$
	0.0065	251.5	103.4	0.022
	$\sigma_s = 220$ MPa			
	$\varepsilon_0/(\%)$	$E_p/(\text{GPa})$	$\sigma_{tr}/(\text{MPa})$	$q/(\%)$
	0.02	209.6	143	0.058
	$\sigma_s = 250$ MPa			
	$\varepsilon_0/(\%)$	$E_p/(\text{GPa})$	$\sigma_{tr}/(\text{MPa})$	$q/(\%)$
	0.059	188.5	173.8	0.138
$\sigma_{fc} = 2.5$ GPa	$\sigma_s = 180$ MPa			
	$\varepsilon_0/(\%)$	$E_p/(\text{GPa})$	$\sigma_{tr}/(\text{MPa})$	$q/(\%)$
	0.028	251.5	103.4	0.0087
	$\sigma_s = 220$ MPa			
	$\varepsilon_0/(\%)$	$E_p/(\text{GPa})$	$\sigma_{tr}/(\text{MPa})$	$q/(\%)$
	0.007	209.6	143	0.02
	$\sigma_s = 250$ MPa			
	$\varepsilon_0/(\%)$	$E_p/(\text{GPa})$	$\sigma_{tr}/(\text{MPa})$	$q/(\%)$
	0.015	188.5	173.8	0.036

- Under $\sigma_s = 250$ MPa, the initial composite strain is $\varepsilon_0 = 0.015\%$; the composite initial tangent modulus is approximately $E_p = 188.5$ GPa, the degradation rate of the composite tangent modulus is 40% compared with original specimen, and the fibers broken fraction is $q = 0.036$. With increasing applied stress to $\sigma_{tr} = 173.8$ MPa, the interface debonding ratio increases, and the composite tangent modulus decreases to approximately $E_p = 86$ GPa corresponding to the interface debonding ratio of $\eta = 0.571$. When the applied stress increases from $\sigma_{tr} = 173.8$ MPa to $\sigma = 250$ MPa, the composite tangent modulus remains constant value of approximately $E_p = 100.6$ GPa with $\eta = 0.579$.

1.4 Experimental Comparisons

Li et al. (2014), Liu (2019), Guo and Kagawa (2002), and Morscher (1999) investigated tensile behavior of unidirectional and 2D SiC/SiC composites at room temperature. In this section, using the developed damage models and micromechanical constitutive models for different damage modes, experimental tensile stress–strain curves are predicted. Comparisons between tensile stress–strain curves with and without stochastic loading are analyzed. Relationships between stochastic loading stress levels, composite tangent modulus, interface debonding ratio, and fibers broken fraction are established.

1.4.1 2D SiC/SiC Under Stochastic Loading of 140, 180, 200, and 240 MPa

Li et al. (2014) investigated the tensile behavior of 2D SiC/SiC composite at room temperature. The composite was fabricated using chemical vapor infiltration (CVI) method. Tensile test was performed under displacement control with the speed of 0.3 mm/min. Experimental tensile stress–strain curves, composite tangent modulus versus applied strain curves, interface debonding ratio and broken fibers fraction versus applied stress curves of 2D SiC/SiC composite under $\sigma_s = 140$, 180, 200, and 240 MPa at room temperature are shown in Fig. 1.13 and Table 1.6. When stochastic loading stress increases, the initial composite strain increases, the initial composite tangent modulus decreases, the transition stress for interface debonding increases, and the initial fiber broken fraction increases.

- Under $\sigma_s = 140$ MPa, the initial strain is $\varepsilon_0 = 0.001\%$; the composite initial tangent modulus is approximately $E_p = 288$ GPa, the degradation rate of composite tangent modulus is 7% compared with original specimen, and the fibers broken fraction is $q = 0.006$. With increasing applied stress to $\sigma_{tr} = 63.8$ MPa, the composite tangent modulus decreases to approximately $E_p = 267.4$ GPa corresponding to $\eta = 0.024$. When the applied stress increases from $\sigma_{tr} = 63.8$ MPa to $\sigma = 140$ MPa, the composite tangent modulus remains constant value of approximately $E_p = 282.3$ GPa with $\eta = 0.025$.
- Under $\sigma_s = 180$ MPa, the initial strain is $\varepsilon_0 = 0.005\%$; the composite initial tangent modulus is approximately $E_p = 251$ GPa, and the degradation rate of composite tangent modulus is 19% compared with original specimen, and the

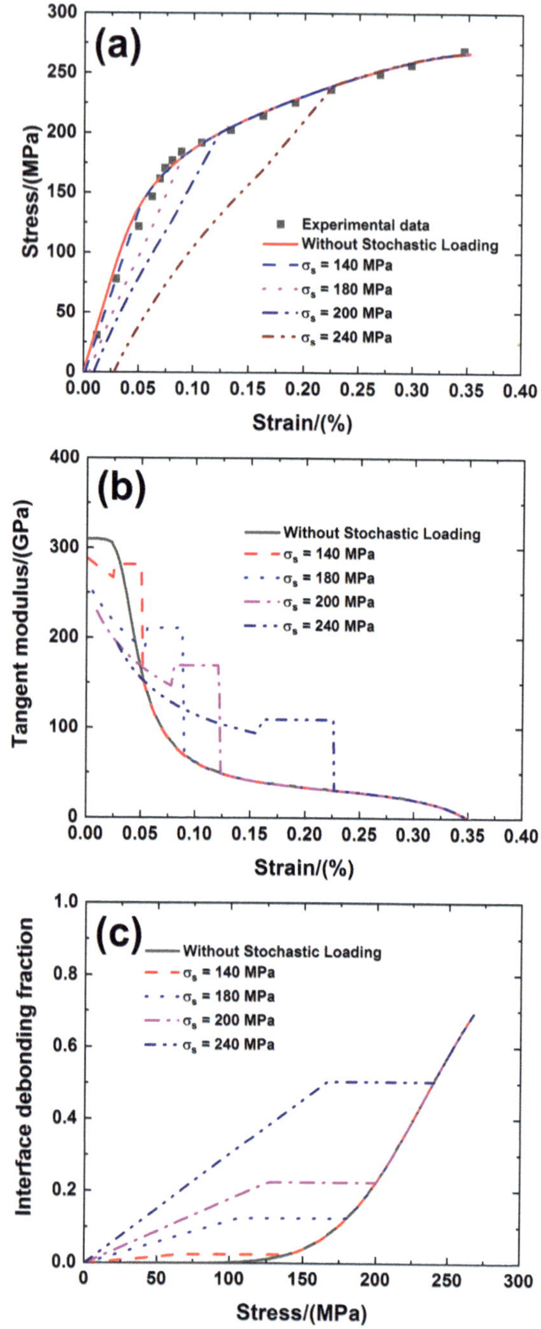

Fig. 1.13 **a** Experimental and predicted ensile stress–strain curves; **b** the predicted composite tangent modulus versus applied strain curves; **c** the predicted interface debonding ratio versus applied stress curves; and **d** the broken fibers fraction versus applied stress curves of 2D SiC/SiC composite under $\sigma_s = 140$, 180, 200, and 240 MPa

Fig. 1.13 (continued)

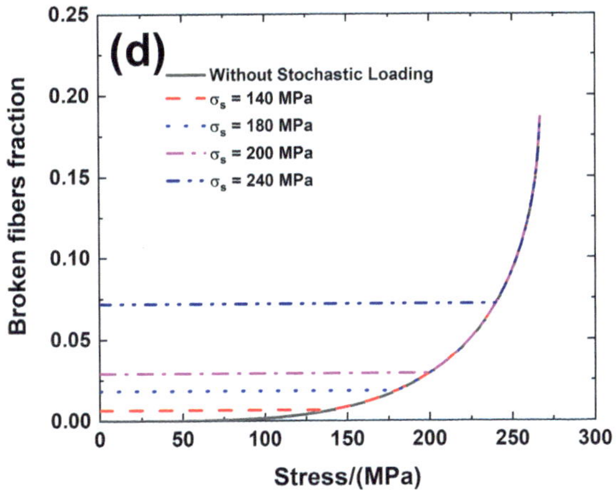

Table 1.6 Tensile damage and fracture of 2D SiC/SiC composite under $\sigma_s = 140$, 180, 200, and 240 MPa

$\sigma_s = 140$ MPa			
ε_0/(%)	E_p/(GPa)	σ_{tr}/(MPa)	q
0.001	288	63.8	0.006
$\sigma_s = 180$ MPa			
ε_0/(%)	E_p/(GPa)	σ_{tr}/(MPa)	q
0.005	251	103.4	0.018
$\sigma_s = 180$ MPa			
ε_0/(%)	E_p/(GPa)	σ_{tr}/(MPa)	q
0.009	229.5	123.2	0.028
$\sigma_s = 180$ MPa			
ε_0/(%)	E_p/(GPa)	σ_{tr}/(MPa)	q
0.028	194.2	162.8	0.078

fibers broken fraction is $q = 0.018$. With increasing applied stress to $\sigma_{tr} = 103.4$ MPa, the composite tangent modulus decreases to approximately $E_p = 187.4$ GPa corresponding to $\eta = 0.122$. When the applied stress increases from $\sigma_{tr} = 103.4$ MPa to $\sigma = 180$ MPa, the composite tangent modulus remains constant value of approximately $E_p = 211.3$ GPa with $\eta = 0.126$.

- Under $\sigma_s = 200$ MPa, the initial composite strain is $\varepsilon_0 = 0.009\%$; the initial composite tangent modulus is approximately $E_p = 229.5$ GPa, the degradation rate of composite tangent modulus is 26% compared with original specimen, and the fibers broken fraction is $q = 0.028$. With increasing applied stress to $\sigma_{tr} = 123.2$ MPa, the composite tangent modulus decreases to $E_p = 147.1$ GPa corresponding to $\eta = 0.22$. When the applied stress increases from $\sigma_{tr} = 123.2$ MPa to $\sigma = 200$ MPa, the composite tangent modulus remains constant value of approximately $E_p = 169.7$ GPa with $\eta = 0.226$.

- Under σ_s = 240 MPa, the initial composite strain is ε_0 = 0.028%; the initial tangent modulus is approximately E_p = 194.2 GPa, the degradation rate of composite tangent modulus is 38% compared with original specimen, and the fibers broken fraction is q = 0.078. With increasing applied stress to σ_{tr} = 162.8 MPa, the composite tangent modulus decreases to approximately E_p = 94 GPa corresponding to η = 0.49. When the applied stress increases from σ_{tr} = 162.8 MPa to σ = 240 MPa, the composite tangent modulus remains constant value of approximately E_p = 109.8 GPa with η = 0.5.

1.4.2 UD and 2D SiC/SiC Under Stochastic Loading

Liu (2019) investigated the tensile behavior of unidirectional and 2D SiC/SiC composites at room temperature. Tensile test was performed under displacement control with the loading rate of 0.2 mm/min.

For unidirectional SiC/SiC composite, the tensile stress–strain curves, composite tangent modulus versus applied strain curves, interface debonding ratio, and broken fibers fraction versus applied stress curves under σ_s = 140, 180, 200, and 220 MPa at room temperature are shown in Fig. 1.14 and Table 1.7.

- Under σ_s = 140 MPa, the initial composite strain is ε_0 = 0.002%; the initial composite tangent modulus is approximately E_p = 235.9 GPa, the degradation rate of composite tangent modulus is 1% compared with original specimen, and the fibers broken fraction is q = 0.01. With increasing applied stress to σ_{tr} = 58.8 MPa, the composite tangent modulus decreases to approximately E_p = 235.7 GPa corresponding to η = 0.0001. When the applied stress increases from σ_{tr} = 58.8 MPa to σ = 140 MPa, the composite tangent modulus remains constant value of approximately E_p = 235.8 GPa with η = 0.0001.
- Under σ_s = 180 MPa, the initial composite strain is ε_0 = 0.009%; the initial composite tangent modulus is E_p = 234 GPa, the degradation rate of composite tangent modulus is 1.1% compared with original specimen, and the fibers broken fraction is q = 0.03. With increasing applied stress to σ_{tr} = 98.4 MPa, the composite tangent modulus decreases to approximately E_p = 230.1 GPa corresponding to η = 0.005. When the applied stress increases from σ_{tr} = 98.4 MPa to σ = 180 MPa, the composite tangent modulus remains constant value of approximately E_p = 230.9 GPa with η = 0.005.
- Under σ_s = 200 MPa, the initial composite strain is ε_0 = 0.0174%; the initial composite tangent modulus is E_p = 229.6 GPa, the degradation rate of composite tangent modulus is 2.8% compared with original specimen, and the fibers broken fraction is q = 0.048. With increasing applied stress to σ_{tr} = 118.8 MPa, the composite tangent modulus decreases to approximately E_p = 215.4 GPa corresponding to η = 0.023. When the applied stress increases from σ_{tr} = 118.8 MPa to σ = 200 MPa, the composite tangent modulus remains constant value of approximately E_p = 217.8 GPa with η = 0.023.

Fig. 1.14 **a** Experimental and predicted tensile stress–strain curves; **b** the predicted composite tangent modulus versus applied strain curves; **c** the predicted interface debonding ratio versus applied stress curves; and **d** the predicted broken fibers fraction versus applied stress curves of UD SiC/SiC composite under $\sigma_s = 140$, 180, 200, and 220 MPa

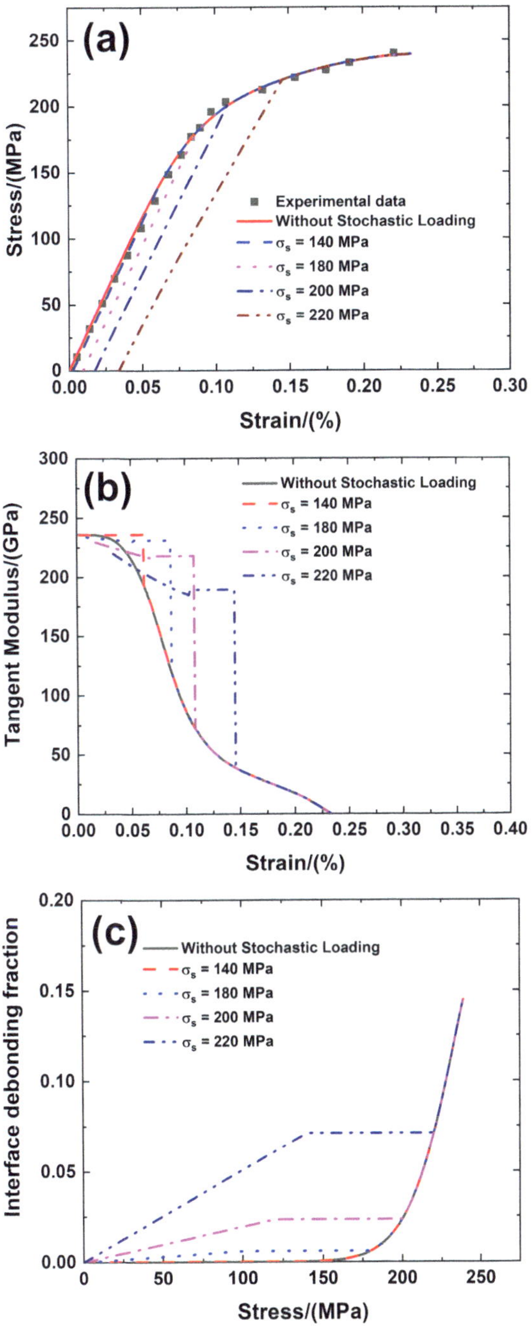

Fig. 1.14 (continued)

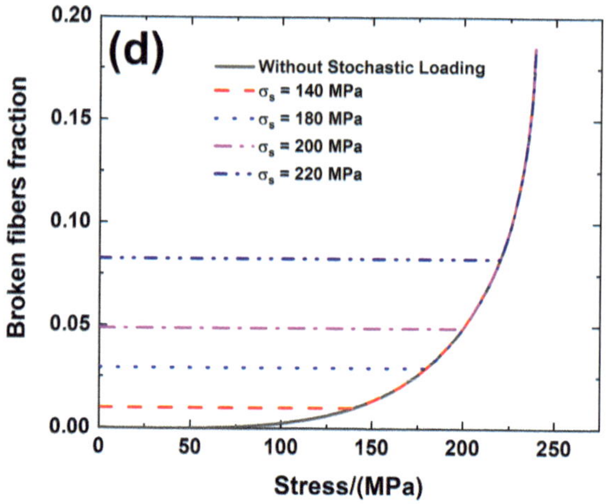

Table 1.7 Tensile damage and fracture of unidirectional SiC/SiC composite under σ_s = 140, 180, 200, and 220 MPa

$\sigma_s = 140$ MPa			
ε_0/(%)	E_p/(GPa)	σ_{tr}/(MPa)	q
0.002	235.9	58.8	0.01
$\sigma_s = 180$ MPa			
ε_0/(%)	E_p/(GPa)	σ_{tr}/(MPa)	q
0.009	234	98.4	0.03
$\sigma_s = 200$ MPa			
ε_0/(%)	E_p/(GPa)	σ_{tr}/(MPa)	q
0.017	229.6	118.8	0.048
$\sigma_s = 220$ MPa			
ε_0/(%)	E_p/(GPa)	σ_{tr}/(MPa)	q
0.033	220.1	139	0.082

- Under $\sigma_s = 220$ MPa, the initial composite strain is $\varepsilon_0 = 0.033\%$; the initial composite tangent modulus is $E_p = 220.1$ GPa, the degradation rate of composite tangent modulus is 6.8% compared with original specimen, and the fibers broken fraction is $q = 0.082$. With increasing applied stress to $\sigma_{tr} = 139$ MPa, the composite tangent modulus decreases to approximately $E_p = 184.5$ GPa corresponding to $\eta = 0.07$. When the applied stress increases from $\sigma_{tr} = 139$ MPa to $\sigma = 220$ MPa, the composite tangent modulus remains constant value of approximately $E_p = 189.2$ GPa with $\eta = 0.071$.

For 2D SiC/SiC composite, the tensile stress–strain curves, composite tangent modulus versus applied strain curves, interface debonding ratio, and broken fibers fraction versus stress curves under $\sigma_s = 80$, 100, and 120 MPa at room temperature are shown in Fig. 1.15 and Table 1.8.

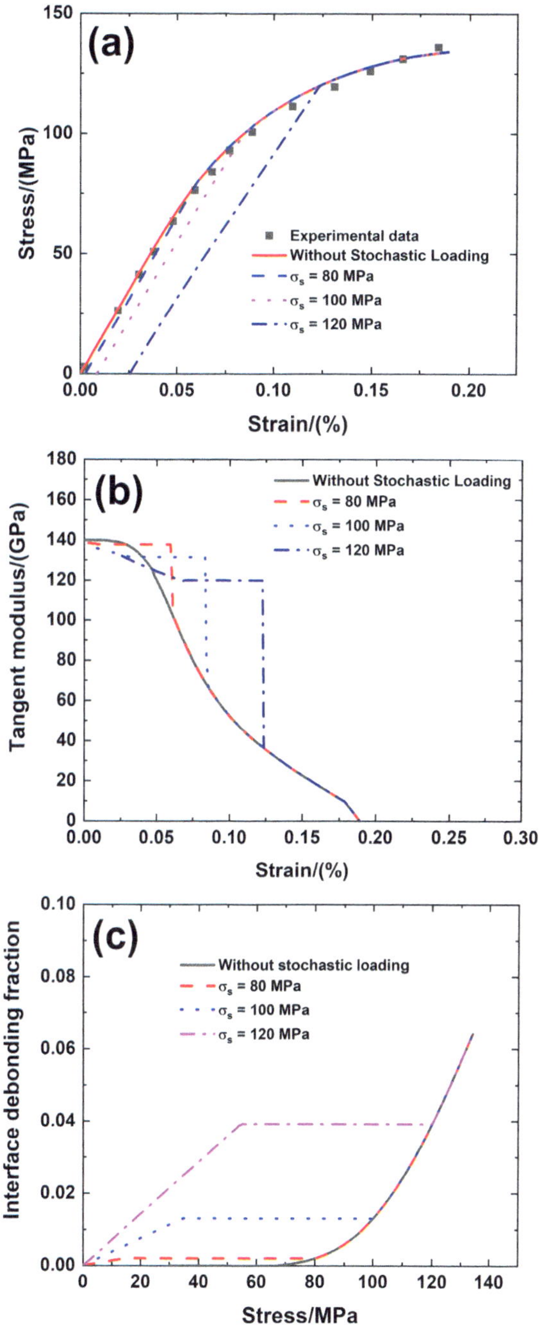

Fig. 1.15 **a** Experimental and predicted tensile stress–strain curves; **b** the predicted composite tangent modulus versus applied strain curves; **c** the predicted interface debonding ratio versus applied stress curves; and **d** the predicted broken fibers fraction versus applied stress curves of 2D SiC/SiC composite under $\sigma_s = 80$, 100, and 120 MPa

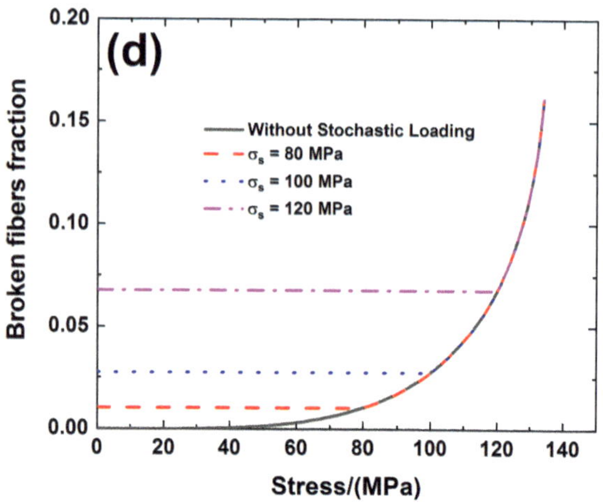

Fig. 1.15 (continued)

Table 1.8 Tensile damage and fracture of 2D SiC/SiC composite under $\sigma_s = 80$, 100, and 120 MPa

$\sigma_s = 80$ MPa			
ε_0/(%)	E_p/(GPa)	σ_{tr}/(MPa)	q
0.002	138.3	13.2	0.01
$\sigma_s = 100$ MPa			
ε_0/(%)	E_p/(GPa)	σ_{tr}/(MPa)	q
0.008	135.6	32.4	0.027
$\sigma_s = 120$ MPa			
ε_0/(%)	E_p/(GPa)	σ_{tr}/(MPa)	q
0.025	131.9	52.8	0.068

- Under $\sigma_s = 80$ MPa, the initial composite strain is $\varepsilon_0 = 0.002\%$; the composite initial tangent modulus is $E_p = 138.3$ GPa, the degradation rate of composite tangent modulus is 1.2% compared with original specimen, and the fibers broken fraction is $q = 0.01$. With increasing applied stress to $\sigma_{tr} = 13.2$ MPa, the composite tangent modulus decreases to approximately $E_p = 137.6$ GPa corresponding to $\eta = 0.001$. When the applied stress increases from $\sigma_{tr} = 13.2$ MPa to $\sigma = 80$ MPa, the composite tangent modulus remains constant value of approximately $E_p = 137.7$ GPa with $\eta = 0.002$.
- Under $\sigma_s = 100$ MPa, the initial composite strain is $\varepsilon_0 = 0.008\%$; the initial composite tangent modulus is approximately $E_p = 135.6$ GPa, the degradation rate of composite tangent modulus is 3.1% compared with original specimen, and the fibers broken fraction is $q = 0.027$. With increasing applied stress to $\sigma_{tr} = 32.4$ MPa, the composite tangent modulus decreases to approximately $E_p = 131$ GPa corresponding to $\eta = 0.012$. When the applied stress increases from $\sigma_{tr} = 32.4$ MPa to $\sigma = 100$ MPa, the composite tangent modulus remains constant value of approximately $E_p = 131.3$ GPa with $\eta = 0.013$.

- Under $\sigma_s = 120$ MPa, the composite initial strain is $\varepsilon_0 = 0.025\%$; the composite initial tangent modulus is $E_p = 131.9$ GPa, the degradation rate of composite tangent modulus is 5.8% compared with original specimen, and the fibers broken fraction is $q = 0.068$. With increasing applied stress to $\sigma_{tr} = 52.8$ MPa, the composite tangent modulus decreases to approximately $E_p = 119.5$ GPa corresponding to $\eta = 0.038$. When the applied stress increases from $\sigma_{tr} = 52.8$ MPa to $\sigma = 120$ MPa, the composite tangent modulus remains constant value of approximately $E_p = 119.5$ GPa with $\eta = 0.039$.

1.4.3 2D SiC/SiC Under Stochastic Loading of 80, 100, and 120 MPa

Guo and Kagawa (2002) investigated the tensile behavior of 2D plain-woven fabric SiC/SiC composite fabricated by the PIP process. The quasi-static tensile test was conducted under displacement control with the rate of 0.5 mm/min. The tensile stress–strain curves, composite tangent modulus versus applied strain curves, interface debonding ratio, and broken fibers fraction versus applied stress curves under $\sigma_s =$ 80, 100, and 120 MPa at room temperature are shown in Fig. 1.16 and Table 1.9.

- Under $\sigma_s = 80$ MPa, the initial composite strain is $\varepsilon_0 = 0.005\%$; the initial composite tangent modulus is approximately $E_p = 56$ GPa, the degradation rate of composite tangent modulus is 6.7% compared with original specimen, and the fibers broken fraction is $q = 0.01$. With increasing applied stress to $\sigma_{tr} =$ 33.6 MPa, the composite tangent modulus decreases to approximately $E_p = 49.4$ GPa corresponding to $\eta = 0.07$. When the applied stress increases from $\sigma_{tr} =$ 33.6 MPa to $\sigma = 80$ MPa, the composite tangent modulus remains constant value of approximately $E_p = 50.4$ GPa with $\eta = 0.074$.
- Under $\sigma_s = 100$ MPa, the initial composite strain is $\varepsilon_0 = 0.016\%$; the initial composite tangent modulus is $E_p = 53.5$ GPa, the degradation rate of composite tangent modulus is 11% compared with original specimen, and the fibers broken fraction is $q = 0.027$. With increasing applied stress to $\sigma_{tr} = 52.8$ MPa, the composite tangent modulus decreases to approximately $E_p = 39.7$ GPa corresponding to $\eta = 0.19$. When the applied stress increases from $\sigma_{tr} = 52.8$ MPa to $\sigma = 100$ MPa, the composite tangent modulus remains constant value of approximately $E_p = 40.7$ GPa with $\eta = 0.2$.
- Under $\sigma_s = 120$ MPa, the initial composite strain is $\varepsilon_0 = 0.046\%$; the initial composite tangent modulus is approximately $E_p = 51$ GPa, the degradation rate of composite tangent modulus is 15% compared with original specimen, and the fibers broken fraction is $q = 0.068$. With increasing applied stress to $\sigma_{tr} =$ 73.2 MPa, the composite tangent modulus decreases to approximately $E_p = 30.8$ GPa corresponding to $\eta = 0.39$. When the applied stress increases from $\sigma_{tr} =$ 73.2 MPa to $\sigma = 120$ MPa, the composite tangent modulus remains constant value of approximately $E_p = 31.7$ GPa with $\eta = 0.395$.

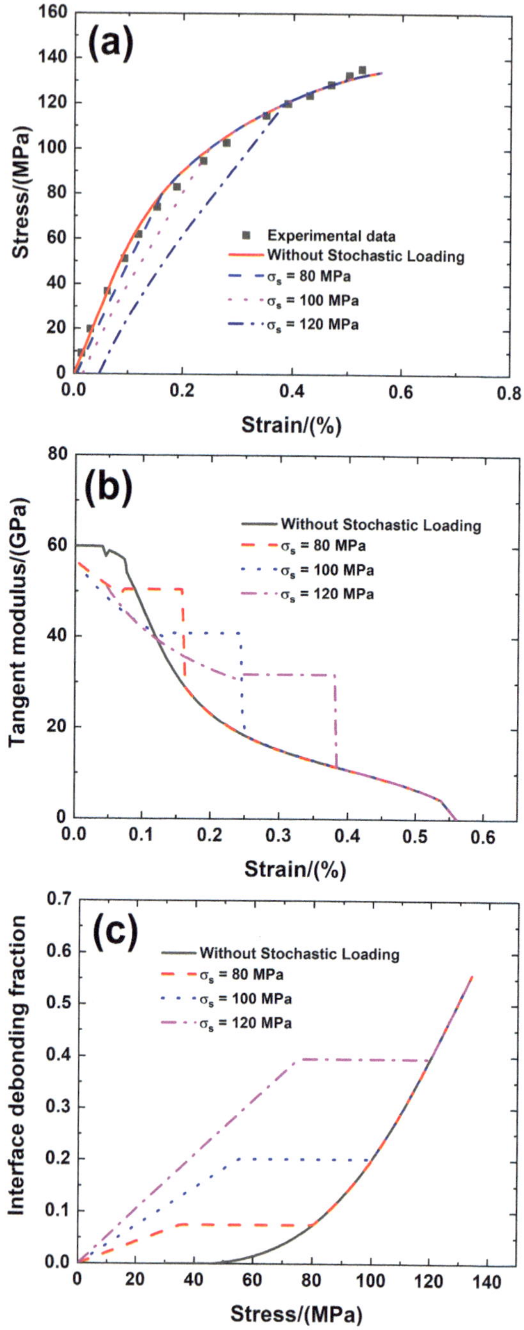

Fig. 1.16 **a** Experimental and predicted tensile stress–strain curves; **b** the predicted composite tangent modulus versus applied strain curves; **c** the predicted interface debonding ratio versus applied stress curves; and **d** the predicted broken fibers fraction versus applied stress curves of 2D SiC/SiC composite under $\sigma_s = 80$, 100, and 120 MPa

Fig. 1.16 (continued)

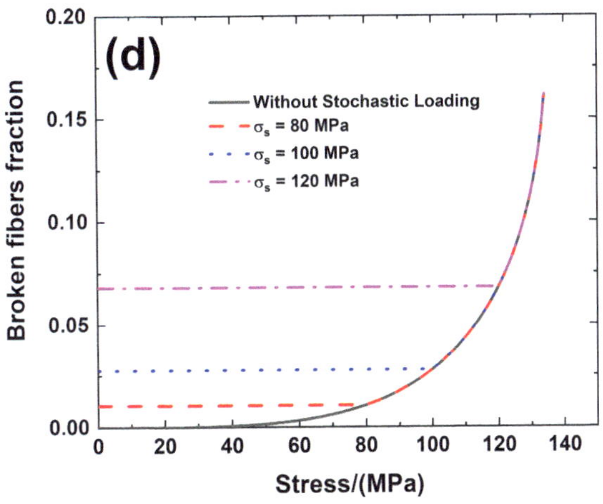

Table 1.9 Tensile damage and fracture of 2D SiC/SiC composite under $\sigma_s = 80$, 100, and 120 MPa

$\sigma_s = 80$ MPa			
$\varepsilon_0/(\%)$	$E_p/(\text{GPa})$	$\sigma_{tr}/(\text{MPa})$	q
0.005	56	33.6	0.01
$\sigma_s = 100$ MPa			
$\varepsilon_0/(\%)$	$E_p/(\text{GPa})$	$\sigma_{tr}/(\text{MPa})$	q
0.016	53.5	52.8	0.027
$\sigma_s = 120$ MPa			
$\varepsilon_0/(\%)$	$E_p/(\text{GPa})$	$\sigma_{tr}/(\text{MPa})$	q
0.046	51	73.2	0.068

1.4.4 2D SiC/SiC Under Stochastic Loading of 180, 220, 260, and 300 MPa

Morscher (1999) investigated tensile behavior of 2D SiC/SiC composite at room temperature. The tensile test was conducted under load control. The tensile stress–strain curves, composite tangent modulus versus applied strain curves, interface debonding ratio, and broken fibers fraction versus applied stress curves under $\sigma_s =$ 180, 220, 260, and 300 MPa are shown in Fig. 1.17 and Table 1.10.

- Under $\sigma_s = 180$ MPa, the initial composite strain is $\varepsilon_0 = 0.003\%$; the initial composite tangent modulus is $E_p = 142.3$ GPa, the degradation rate of composite tangent modulus is 50% compared with original specimen, and the fibers broken fraction is $q = 0.004$. With increasing applied stress to $\sigma_{tr} = 122$ MPa, the composite tangent modulus decreases to approximately $E_p = 53.9$ GPa corresponding to $\eta = 0.63$. When the applied stress increases from $\sigma_{tr} = 122$ MPa

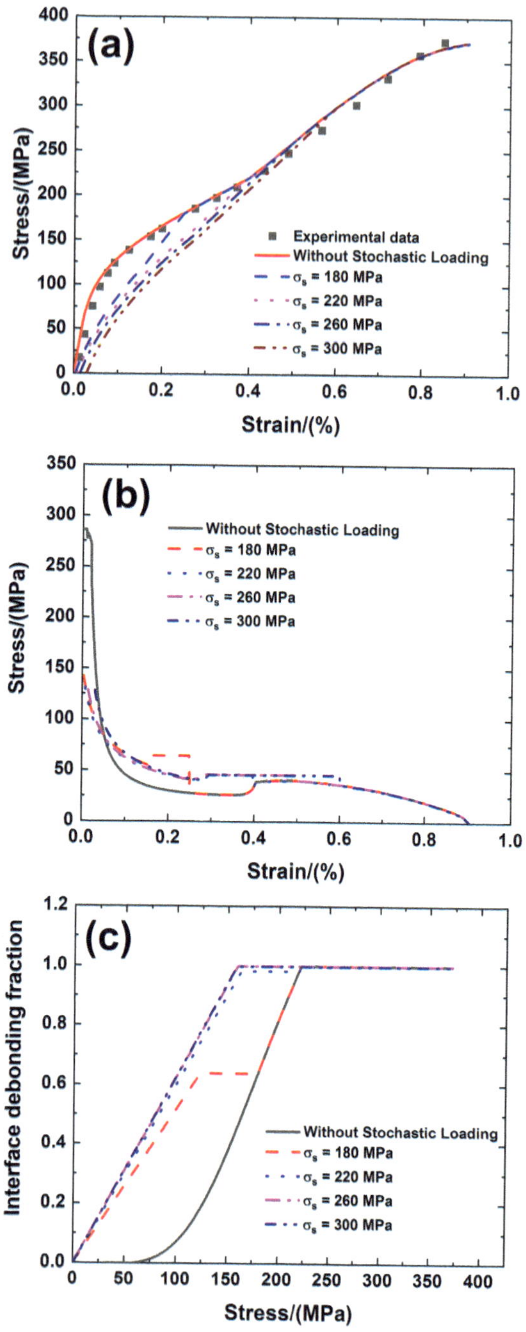

Fig. 1.17 **a** Experimental and predicted tensile stress–strain curves; **b** the predicted composite tangent modulus versus applied strain curves; **c** the predicted interface debonding ratio versus applied stress curves; and **d** the broken fibers fraction versus applied stress curves of 2D SiC/SiC composite under $\sigma_s = 180$, 220, 260, and 300 MPa

Fig. 1.17 (continued)

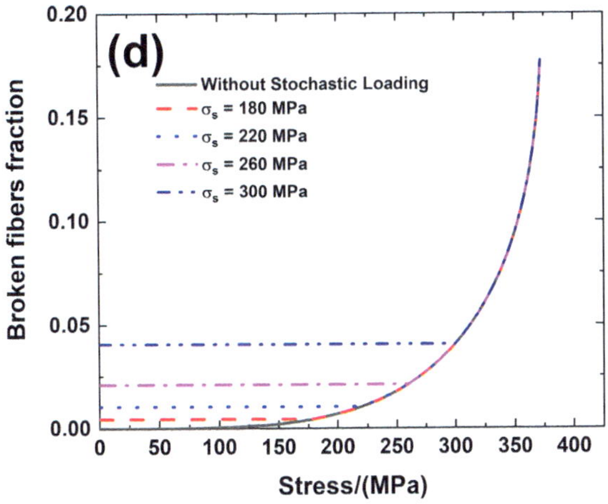

Table 1.10 Tensile damage and fracture of 2D SiC/SiC composite under $\sigma_s = 180$, 220, 260, and 300 MPa

$\sigma_s = 180$ MPa			
$\varepsilon_0/(\%)$	$E_p/(\mathrm{GPa})$	$\sigma_{tr}/(\mathrm{MPa})$	q
0.003	142.3	122	0.004
$\sigma_s = 220$ MPa			
$\varepsilon_0/(\%)$	$E_p/(\mathrm{GPa})$	$\sigma_{tr}/(\mathrm{MPa})$	q
0.007	131.4	162	0.01
$\sigma_s = 260$ MPa			
$\varepsilon_0/(\%)$	$E_p/(\mathrm{GPa})$	$\sigma_{tr}/(\mathrm{MPa})$	q
0.014	128.7	150	0.021
$\sigma_s = 300$ MPa			
$\varepsilon_0/(\%)$	$E_p/(\mathrm{GPa})$	$\sigma_{tr}/(\mathrm{MPa})$	q
0.029	128.3	148	0.04

to $\sigma = 180$ MPa, the composite tangent modulus remains constant value of approximately $E_p = 63.9$ GPa with $\eta = 0.638$.

- Under $\sigma_s = 220$ MPa, the initial composite strain is $\varepsilon_0 = 0.007\%$; the initial composite tangent modulus is approximately $E_p = 131.4$ GPa, the degradation rate of composite tangent modulus is 54% compared with original specimen, and the fibers broken fraction is $q = 0.01$. With increasing applied stress to $\sigma_{tr} = 162$ MPa, the composite tangent modulus decreases to approximately $E_p = 42$ GPa corresponding to $\eta = 0.63$. When the applied stress increases from $\sigma_{tr} = 122$ MPa to $\sigma = 180$ MPa, the composite tangent modulus remains constant value of approximately $E_p = 63.9$ GPa with $\eta = 0.97$.
- Under $\sigma_s = 260$ MPa, the initial composite strain is $\varepsilon_0 = 0.014\%$; the initial composite tangent modulus is $E_p = 128.7$ GPa, the degradation rate of composite tangent modulus is 55% compared with original specimen, and the fibers broken

fraction is $q = 0.021$. With increasing applied stress to $\sigma_{tr} = 150$ MPa, the composite tangent modulus decreases to approximately $E_p = 41.3$ GPa corresponding to $\eta = 0.93$. When the applied stress increases from $\sigma_{tr} = 150$ MPa to $\sigma = 260$ MPa, the composite tangent modulus remains constant value of approximately $E_p = 46$ GPa with $\eta = 1.0$.

- Under $\sigma_s = 300$ MPa, the initial composite strain is $\varepsilon_0 = 0.029\%$; the initial composite tangent modulus is $E_p = 128.3$ GPa, the degradation rate of composite tangent modulus is 55.1% compared with original specimen, and the fibers broken fraction is $q = 0.04$. With increasing applied stress to $\sigma_{tr} = 148$ MPa, the composite tangent modulus decreases to approximately $E_p = 41.3$ GPa corresponding to $\eta = 0.93$. When the applied stress increases from $\sigma_{tr} = 148$ MPa to $\sigma = 300$ MPa, the composite tangent modulus remains constant value of approximately $E_p = 46$ GPa with $\eta = 1.0$.

1.5 Conclusions

In this chapter, effect of stochastic loading on tensile damage and fracture of fiber-reinforced CMCs was investigated. A micromechanical constitutive model considering stochastic loading was developed considering multiple damage mechanisms under tensile loading. Relationships between stochastic loading stress, composite tangent modulus, interface debonding ratio, and fibers broken fraction were established. Effects of composite constituent properties and damage state on tensile damage and fracture were analyzed. Experimental tensile damage and fracture of unidirectional and 2D SiC/SiC composites subjected to different stochastic loading stress were predicted.

- When the fiber volume increased, the initial composite strain decreased, the initial composite tangent modulus increased, the transition stress for the interface debonding decreased, and the initial fibers broken fraction decreased.
- When the interface shear stress increased, the initial composite strain remained the same, the initial composite tangent modulus increased, the transition stress for the interface debonding remained the same, and the initial fibers broken fraction remained the same.
- When the interface debonding energy increased, the initial composite strain, composite tangent modulus, and broken fibers fraction remained the same, and the transition stress for the interface debonding decreased.
- When the saturation matrix crack spacing increased, the initial composite strain decreased, the initial tangent modulus increased, and the transition stress for interface debonding and initial fiber broken fraction remained the same.
- When the fiber strength increased, the initial composite strain and fibers broken fraction decreased, and the initial composite tangent composite modulus and transition stress for interface debonding remained the same.

References

Corman GS, Luthra KL (2004) Silicon melt infiltrated ceramic composites (HiPerComp^{TM}). In: Bansal NP (ed) Handbook of ceramic composites. Kluwer Academic Publisher, Boston

Curtin WA (1991) Theory of mechanical properties of ceramic-matrix composites. J Am Ceram Soc 74:2837–2845. https://doi.org/10.1111/j.1151-2916.1991.tb06852.x

Curtin WA (1993) Multiple matrix cracking in brittle matrix composites. Acta Metal Mater 41:1369–1377. https://doi.org/10.1016/0956-7151(93)90246-O

Domergue JC, Vagaggini E, Evans AG (1995) Relationships between hysteresis measurements and the constituent properties of ceramic matrix composites: II, Experimental studies on unidirectional materials. J Am Ceram Soc 78:2721–2731. https://doi.org/10.1111/j.1151-2916.1995.tb08047.x

Evans AG (1997) Design and life prediction issues for high-temperature engineering ceramics and their composites. Acta Mater 45:23–40. https://doi.org/10.1016/S1359-6454(96)00143-7

Goto K, Kagawa Y (1996) Fracture behavior and toughness of a plain-woven SiC fiber-reinforced SiC matrix composite. Mater Sci Eng A 211:72–81. https://doi.org/10.1016/0921-5093(96)10170-2

Guo S, Kagawa Y (2002) Tensile fracture behavior of continuous SiC fiber-reinforced SiC matrix composites at elevated temperatures and correlation to in situ constituent properties. J Eur Ceram Soc 22:2349–2356. https://doi.org/10.1016/S0955-2219(02)00028-6

Guo SQ, Kagawa Y, Tanaka Y, Masuda C (1998) Microstructure and role of outermost coating for tensile strength of SiC fiber. Acta Mater 46:4941–4954. https://doi.org/10.1016/S1359-6454(98)00181-5

Li LB (2014) Assessment of the interfacial properties from fatigue hysteresis loss energy in ceramic-matrix composites with different fiber preforms at room and elevated temperatures. Mater Sci Eng A 613:17–36. https://doi.org/10.1016/j.msea.2014.06.092

Li LB (2018) Modeling the monotonic and cyclic tensile stress-strain behavior of 2D and 2.5D woven C/SiC ceramic-matrix composites. Mech Compos Mater 54:165–178. https://doi.org/10.1007/s11029-018-9729-5

Li LB (2020b) Modeling matrix fracture in fiber-reinforced ceramic-matrix composites with different fiber preforms. Text Res J 90:909–924. https://doi.org/10.1177/0040517519883956

Li LB (2020a) Effect of stochastic loading on tensile damage and fracture of fiber-reinforced ceramic-matrix composites. Materials 13:2469. https://doi.org/10.3390/ma13112469

Liu Y (2019) Study on damage failure mechanism of SiC/SiC ceramic matrix composites. Master thesis, Harbin Institute of Technology

Liu H, Yang J, Chen Z, Lang X, Jiao J (2020) Microstructure and properties of SiC_f/SiC composite fabricated by melt infiltration process. Aerospace Mater Technol 6:1–7

Li LB, Song Y, Sun Y (2013) Modeling the tensile behavior of unidirectional C/SiC ceramic-matrix composites. Mech Compos Mater 49:659–672. https://doi.org/10.1007/s11029-013-9382-y

Li P, Wang B, Zhen W, Jiao G (2014) Tensile loading/unloading stress-strain behavior of 2D-SiC/SiC composites. Acta Materiae Compositae Sinica 31:676–682

Li LB, Song Y, Sun Y (2015) Modeling the tensile behavior of cross-ply C/SiC ceramic-matrix composites. Mech Compos Mater 51:359–376. https://doi.org/10.1007/s11029-015-9507-6

Li LB, Reynaud P, Fantozzi G (2020) Cyclic-dependent damage evolution in self-healing woven SiC/[Si-B-C] ceramic-matrix composites at elevated temperatures. Materials 13:1478. https://doi.org/10.3390/ma13061478

Marshall DB, Cox BN, Evans AG (1985) The mechanics of matrix cracking in brittle-matrix fiber composites. Acta Metal 33:2013–2021. https://doi.org/10.1016/0001-6160(85)90124-5

McNulty JC, Zok FW (1999) Low-cycle fatigue of Nicalon-fiber-reinforced ceramic composites. Compos Sci Technol 59:1597–1607. https://doi.org/10.1016/S0266-3538(99)00019-6

Meyer P, Wass AM (2018) Experimental results on the elevated temperature tensile response of SiC/SiC ceramic matrix notched composites. Compos Part B 143:269–281. https://doi.org/10.1016/j.compositesb.2018.01.019

Morscher GN (1999) Modal acoustic emission of damage accumulation in a woven SiC/SiC composite. Compos Sci Technol 59:687–697. https://doi.org/10.1016/S0266-3538(98)00121-3

Morscher GN, Pujar VV (2009) Design guidance for in-plane mechanical properties of SiC fiber-reinforced melt-infiltrated SiC composites. Int J Appl Ceram Technol 6:151–163. https://doi.org/10.1111/j.1744-7402.2008.02331.x

Morscher GN, Yun HM, DiCarlo JA (2007) In-plane cracking behavior and ultimate strength for 2D woven and braided melt-infiltrated SiC/SiC composites tensile loaded in off-axis fiber directions. J Am Ceram Soc 90:3185–3193. https://doi.org/10.1111/j.1551-2916.2007.01887.x

Naslain R (2004) Design, preparation and properties of non-oxide CMCs for application in engines and nuclear reactors: an overview. Compos Sci Technol 64:155–170. https://doi.org/10.1016/S0266-3538(03)00230-6

Naslain R, Lamon J, Pailler R, Bourrat X, Guette A, Langlais F (1999) Micro/minicomposites: a useful approach to the design and development of non-oxide CMCs. Compos Part A 30:537–547. https://doi.org/10.1016/S1359-835X(98)00147-X

Ruggles-Wrenn MB, Christensen DT, Chamberlain AL, Lane JE, Cook TS (2011) Effect of frequency and environment on fatigue behavior of a CVI SiC/SiC ceramic matrix composite at 1200 °C. Compos Sci Technol 71:190–196. https://doi.org/10.1016/j.compscitech.2010.11.008

Udayakumar A, Ganesh AS, Raja S. Balasubramanian M (2011) Effect of intermediate heat treatment on mechanical properties of SiC_f/SiC composites with BN interphase prepared by ICVI. J Eur Ceram Soc 31:1145−1153. https://doi.org/10.1016/j.jeurceramsoc.2010.12.018

Yin J, Lee SH, Feng L, Zhu Y, Liu X, Huang Z, Kim S-Y, Han I-S (2015) The effects of SiC precursors on the microstructures and mechanical properties of SiC_f/SiC composites prepared via polymer impregnation and pyrolysis process. Ceram Int 41:4145–4153. https://doi.org/10.1016/j.ceramint.2014.11.112

Zok FW, Spearing SM (1992) Matrix crack spacing in brittle matrix composites. Acta Metal Mater 40:2033–2043. https://doi.org/10.1016/0956-7151(92)90189-L

Chapter 2
Hysteresis Loops of Ceramic-Matrix Composites Subjected to Stochastic Loading

Abstract In this chapter, synergistic effects of stochastic loading sequence and interface wear on fatigue hysteresis loops of fiber-reinforced ceramic-matrix composites (CMCs) are investigated. Based on the fatigue damage mechanisms of fiber sliding relative to matrix in the interface debonding region, the interface debonding length, unloading interface counter-slip length, and reloading interface new-slip length are determined using fracture mechanics approach. Effects of stochastic stress level, material properties, interface wear, and stochastic loading sequence on the interface slip and fatigue hysteresis loops are analyzed.

Keywords Ceramic-matrix composites (CMCs) · Hysteresis loops · Stochastic loading

2.1 Introduction

Ceramic-matrix composites (CMCs) have the characteristics of low density, high-temperature resistance, wear resistance, and long life, which mainly include C/SiC and SiC/SiC composites (Naslain 2004; Li 2018, 2019, 2020). The density of SiC/SiC composite is only approximately 1/4 – 1/3 of that of nickel-base superalloy, but the operating temperature is approximately 400–500 °C higher than that of the nickel-base superalloy. The development of CMC components follows the idea of "easy first, then difficult," that is, from static parts to rotating parts, from low temperature to high temperature. Priority has been given to the development of stationary parts with intermediate temperature (i.e., 700–1000 °C) and medium load (i.e., lower than 120 MPa), for example, nozzle flap seal and regulator and inner cone, and then to develop stationary parts with high-temperature (i.e., 1000–1300 °C) and medium load, such as flame tube, flame stabilizer, turbine outer ring, and guide vane, and finally to develop high-pressure turbine rotor parts with higher load (i.e., higher than 120 MPa). At present, the entire life test of nozzle flap seal and regulator, inner cone, and turbine outer ring has already been completed; the life assessment of flame tube and guide vane is in progress; the development of high-pressure turbine rotor is in the verification stage. NASA, GE, and other companies have developed C/SiC turbine blisk, SiC/SiC turbine blades, and other rotor parts, and carried out preliminary tests

L. Li, *Damage and Fracture of Ceramic-Matrix Composites Under Stochastic Loading*, Advanced Ceramics and Composites 2,
https://doi.org/10.1007/978-981-16-2141-3_2

on F110, F414, F136 aeroengine platforms. Turbine rotor is the most challenging key component in the hot section of turbine engine, and the success application of CMCs on turbine rotor will have a significant impact on the development of turbine engine (Li 2019b,c; Wang et al. 2008).

Under cyclic fatigue loading of fiber-reinforced CMCs, matrix multicracking and fiber/matrix interface debonding occur first, and the open and closure of matrix cracking upon each cycle are the basic fatigue damage mechanisms (Evans et al. 1995). The stress–strain hysteresis loops develop due to the frictional slip occurred along any interface debonding region upon unloading and subsequent reloading (Holmes and Cho 1992; Reynaud 1996). The shape, location, and area of the fatigue hysteresis loops can be used as an effective tool to monitor damage evolution of fiber-reinforced CMCs under cyclic fatigue loading (Fantozzi and Reynaud 2009). Many researchers investigated fatigue hysteresis loops models of fiber-reinforced CMCs (Kotil et al. 1990; Pryce and Smith 1993; Ahn and Curtin 1997; Solti et al. 1995,1997; Vagaggini et al. 1995). Li et al. investigated the effect of interface debonding (Li et al. 2009a), fibers Poisson contraction (Li et al. 2009b), fibers fracture (Li and Song 2011; Li 2014), matrix multicracking (Li 2013a,b), and matrix cracking modes (Li 2015a,b) on cyclic loading/unloading and fatigue hysteresis loops of unidirectional, cross-ply, and 2.5D woven CMCs under single fatigue stress level. Li (2015c) investigated the effect of single/multiple loading stress levels and different loading sequences on hysteresis loops of fiber-reinforced CMCs. During applications of ceramic composite structures on aircraft thermo-structural components, there would exist synergistic effect of stochastic loading and interface wear mechanism on fatigue performance, in which the interface shear stress in the interface wear region under low fatigue peak stress level is much lower than that in the new interface debonding region under high fatigue peak stress level. The difference of interface shear stress existing in the interface wear region and new interface debonding region affects the interface slip lengths, then affect the shape, location, and area of fatigue hysteresis loops (Li 2015d).

In this chapter, synergistic effects of stochastic loading and interface wear on fatigue hysteresis loops of fiber-reinforced CMCs are investigated. Based on the fatigue damage mechanisms of fiber sliding relative to matrix upon unloading and subsequent reloading and considering the difference of interface shear stress existed in the new and original interface debonding region, the interface debonding length, unloading interface counter-slip length, and reloading interface new-slip length are determined using the fracture mechanics approach. The fatigue hysteresis loops models considering stochastic loading and interface wear mechanism are developed. Effects of fatigue peak stress, fiber volume, matrix crack spacing, interface debonding and interface wear, and stochastic loading stress levels on interface frictional slip and fatigue hysteresis loops are analyzed.

2.2 Theoretical Models

2.2.1 *Stress Analysis*

Upon first loading to fatigue peak stress $\sigma_{\max1}$, it is assumed that matrix multiple cracking and fiber/matrix interface debonding occur. After experiencing N cycles, the interface shear stress in the interface debonding region decreases from initial value τ_i to τ_f due to interface wear. When the fatigue peak stress increases from $\sigma_{\max1}$ to $\sigma_{\max2}$, the crack propagates along the interface. The microstress field of the damaged composite can be determined as,

$$\sigma_f(x) = \begin{cases} \frac{\sigma}{V_f} - \frac{2\tau_f}{r_f}x, x \in [0, \xi] \\ \frac{\sigma}{V_f} - \frac{2\tau_f}{r_f}\xi - \frac{2\tau_i}{r_f}(x - \xi), x \in [\xi, l_d] \\ \sigma_{fo} + \left[\frac{V_m}{V_f}\sigma_{mo} - 2\frac{\tau_f}{r_f}\xi - 2\frac{\tau_i}{r_f}(l_d - \xi)\right]\exp\left(-\rho\frac{x-l_d}{r_f}\right), x \in \left[l_d, \frac{l_c}{2}\right] \end{cases} \tag{2.1}$$

$$\sigma_m(x) = \begin{cases} 2\frac{V_f}{V_m}\frac{\tau_f}{r_f}x, x \in [0, \xi] \\ 2\frac{V_f}{V_m}\frac{\tau_f}{r_f}\xi + 2\frac{V_f}{V_m}\frac{\tau_i}{r_f}(x - \xi), x \in [\xi, l_d] \\ \sigma_{mo} - \left[\sigma_{mo} - 2\frac{V_f}{V_m}\frac{\tau_f}{r_f}\xi - 2\frac{V_f}{V_m}\frac{\tau_i}{r_f}(l_d - \xi)\right]\exp\left(-\rho\frac{x-l_d}{r_f}\right), x \in \left[l_d, \frac{l_c}{2}\right] \end{cases} \tag{2.2}$$

$$\tau_i(x) = \begin{cases} \tau_f, x \in [0, \xi] \\ \tau_i, x \in [\xi, l_d] \\ \frac{\rho}{2}\left[\frac{V_m}{V_f}\sigma_{mo} - 2\frac{\tau_f}{r_f}\xi - \frac{2\tau_i}{r_f}(l_d - \xi)\right]\exp\left(-\rho\frac{x-l_d}{r_f}\right), x \in \left[l_d, \frac{l_c}{2}\right] \end{cases} \tag{2.3}$$

where V_f and V_m denote the matrix and fiber volume fraction, r_f is the fiber radius, τ_f and τ_i are the interface shear stress in the interface wear region and interface slip region, ξ, l_d, and l_c are the interface wear length, interface debonding length, and matrix crack spacing, ρ denotes the shear-lag model parameter (Budiansky et al. 1986), and σ_{fo} and σ_{mo} denote the fiber and matrix axial stress in the interface bonded region, respectively.

$$\sigma_{fo} = \frac{E_f}{E_c}\sigma + E_f(\alpha_c - \alpha_f)\Delta T \tag{2.4}$$

$$\sigma_{mo} = \frac{E_m}{E_c}\sigma + E_m(\alpha_c - \alpha_m)\Delta T \tag{2.5}$$

where E_f, E_m, and E_c denote the fiber, matrix, and composite elastic modulus, respectively; α_f, α_m, and α_c denote the fiber, matrix, and composite thermal expansion coefficient, respectively; ΔT denotes the temperature difference between fabricated temperature T_0 and testing temperature T_1 ($\Delta T = T_1 - T_0$).

Upon unloading from the fatigue peak stress $\sigma_{\max2}$, the unit cell of half matrix crack spacing can be divided into two regions, i.e., interface debonding region and interface bonding region. The interface debonding region can be divided into three regions, i.e., interface counter-slip region with low interface shear stress τ_{f} ($x \in [0, \xi]$), the interface slip region with high interface shear stress τ_{i} ($x \in [\xi, l_y]$), and the interface slip region with high interface shear stress τ_{i} ($x \in [l_y, l_{\mathrm{d}}]$). The fiber axial stress distribution upon unloading is determined by,

$$\sigma_{\mathrm{f}}(x) = \begin{cases} \frac{\sigma}{V_{\mathrm{f}}} + \frac{2\tau_{\mathrm{f}}}{r_{\mathrm{f}}}x, x \in [0, \xi] \\ \frac{\sigma}{V_{\mathrm{f}}} + \frac{2\tau_{\mathrm{f}}}{r_{\mathrm{f}}}\xi + \frac{2\tau_{\mathrm{i}}}{r_{\mathrm{f}}}(x - \xi), x \in \left[\xi, l_y\right] \\ \frac{\sigma}{V_{\mathrm{f}}} + \frac{2\tau_{\mathrm{f}}}{r_{\mathrm{f}}}\xi + \frac{2\tau_{\mathrm{i}}}{r_{\mathrm{f}}}\left(2l_y - \xi - x\right), x \in \left[l_y, l_{\mathrm{d}}\right] \\ \sigma_{\mathrm{fo}} + \left[\frac{V_{\mathrm{m}}}{V_{\mathrm{f}}}\sigma_{\mathrm{mo}} + \frac{2\tau_{\mathrm{f}}}{r_{\mathrm{f}}}\xi + \frac{2\tau_{\mathrm{i}}}{r_{\mathrm{f}}}\left(2l_y - \xi - l_{\mathrm{d}}\right)\right]\exp\left(-\rho\frac{x - l_{\mathrm{d}}}{r_{\mathrm{f}}}\right), x \in \left[l_{\mathrm{d}}, \frac{l_{\mathrm{c}}}{2}\right] \end{cases} \tag{2.6}$$

The matrix axial stress distribution upon unloading is,

$$\sigma_{\mathrm{m}}(x) = \begin{cases} -2\frac{V_{\mathrm{f}}}{V_{\mathrm{m}}}\frac{\tau_{\mathrm{f}}}{r_{\mathrm{f}}}x, x \in [0, \xi] \\ -2\frac{V_{\mathrm{f}}}{V_{\mathrm{m}}}\frac{\tau_{\mathrm{f}}}{r_{\mathrm{f}}}\xi - 2\frac{V_{\mathrm{f}}}{V_{\mathrm{m}}}\frac{\tau_{\mathrm{i}}}{r_{\mathrm{f}}}(x - \xi), x \in \left[\xi, l_y\right] \\ -2\frac{V_{\mathrm{f}}}{V_{\mathrm{m}}}\frac{\tau_{\mathrm{f}}}{r_{\mathrm{f}}}\xi - 2\frac{V_{\mathrm{f}}}{V_{\mathrm{m}}}\frac{\tau_{\mathrm{i}}}{r_{\mathrm{f}}}\left(2l_y - \xi - x\right), x \in [\xi, l_{\mathrm{d}}] \\ \sigma_{\mathrm{mo}} - \left[\sigma_{\mathrm{mo}} + 2\frac{V_{\mathrm{f}}}{V_{\mathrm{m}}}\frac{\tau_{\mathrm{f}}}{r_{\mathrm{f}}}\xi + 2\frac{V_{\mathrm{f}}}{V_{\mathrm{m}}}\frac{\tau_{\mathrm{i}}}{r_{\mathrm{f}}}\left(2l_y - \xi - l_{\mathrm{d}}\right)\right]\exp\left(-\rho\frac{x - l_{\mathrm{d}}}{r_{\mathrm{f}}}\right), x \in \left[l_{\mathrm{d}}, \frac{l_{\mathrm{c}}}{2}\right] \end{cases} \tag{2.7}$$

The interface shear stress distribution upon unloading is,

$$\tau_{\mathrm{i}}(x) = \begin{cases} -\tau_{\mathrm{f}}, x \in [0, \xi] \\ -\tau_{\mathrm{i}}, x \in \left[\xi, l_y\right] \\ \tau_{\mathrm{i}}, x \in \left[l_y, l_{\mathrm{d}}\right] \\ \frac{\rho}{2}\left[\frac{V_{\mathrm{m}}}{V_{\mathrm{f}}}\sigma_{\mathrm{mo}} + \frac{2\tau_{\mathrm{f}}}{r_{\mathrm{f}}}\xi + \frac{2\tau_{\mathrm{i}}}{r_{\mathrm{f}}}\left(2l_y - \xi - l_{\mathrm{d}}\right)\right]\exp\left(-\rho\frac{x - l_{\mathrm{d}}}{r_{\mathrm{f}}}\right), x \in \left[l_{\mathrm{d}}, \frac{l_{\mathrm{c}}}{2}\right] \end{cases} \tag{2.8}$$

Upon reloading, the new interface slip again occurs near the matrix crack plane over a distance l_z, which is denoted to be interface new-slip region. The interface debonding region can be divided into four regions, i.e., the interface new-slip region with low interface shear stress τ_{f} ($x \in [0, l_z]$), the interface counter-slip region with low interface shear stress τ_{f} ($x \in [l_z, \xi]$), the interface counter-slip region with high interface shear stress τ_{i} ($x \in [\xi, l_y]$), and the interface slip region with high interface shear stress τ_{i} ($x \in [l_y, l_{\mathrm{d}}]$). The fiber axial stress distribution upon reloading is,

$$\sigma_{\mathrm{f}}(x)=\begin{cases}\frac{\sigma}{V_{\mathrm{f}}}-\frac{2\tau_{\mathrm{f}}}{r_{\mathrm{f}}}x,\ x\in[0,l_z]\\ \frac{\sigma}{V_{\mathrm{f}}}-\frac{2\tau_{\mathrm{f}}}{r_{\mathrm{f}}}(2l_z-x),\ x\in[l_z,\xi]\\ \frac{\sigma}{V_{\mathrm{f}}}-\frac{2\tau_{\mathrm{f}}}{r_{\mathrm{f}}}(2l_z-\xi)+\frac{2\tau_{\mathrm{i}}}{r_{\mathrm{f}}}(x-\xi),\ x\in[\xi,l_y]\\ \frac{\sigma}{V_{\mathrm{f}}}-\frac{2\tau_{\mathrm{f}}}{r_{\mathrm{f}}}(2l_z-\xi)+\frac{2\tau_{\mathrm{i}}}{r_{\mathrm{f}}}(2l_y-\xi-x),\ x\in[l_y,l_{\mathrm{d}}]\\ \sigma_{\mathrm{fo}}+\left[\frac{V_{\mathrm{m}}}{V_{\mathrm{f}}}\sigma_{\mathrm{mo}}-\frac{2\tau_{\mathrm{f}}}{r_{\mathrm{f}}}(2l_z-\xi)\right.\\ \left.\quad+\frac{2\tau_{\mathrm{i}}}{r_{\mathrm{f}}}(2l_y-\xi-l_{\mathrm{d}})\right]\exp\left(-\rho\frac{x-l_{\mathrm{d}}}{r_{\mathrm{f}}}\right),\ x\in\left[l_{\mathrm{d}},\frac{l_{\mathrm{c}}}{2}\right]\end{cases}\tag{2.9}$$

The matrix axial stress distribution upon reloading is,

$$\sigma_{\mathrm{m}}(x)=\begin{cases}2\frac{V_{\mathrm{f}}}{V_{\mathrm{m}}}\frac{\tau_{\mathrm{f}}}{r_{\mathrm{f}}}x,\ x\in[0,l_z]\\ 2\frac{V_{\mathrm{f}}}{V_{\mathrm{m}}}\frac{\tau_{\mathrm{f}}}{r_{\mathrm{f}}}(2l_z-x),\ x\in[l_z,\xi]\\ 2\frac{V_{\mathrm{f}}}{V_{\mathrm{m}}}\frac{\tau_{\mathrm{f}}}{r_{\mathrm{f}}}(2l_z-\xi)-2\frac{V_{\mathrm{f}}}{V_{\mathrm{m}}}\frac{\tau_{\mathrm{i}}}{r_{\mathrm{f}}}(x-\xi),\ x\in[\xi,l_y]\\ 2\frac{V_{\mathrm{f}}}{V_{\mathrm{m}}}\frac{\tau_{\mathrm{f}}}{r_{\mathrm{f}}}(2l_z-\xi)-2\frac{V_{\mathrm{f}}}{V_{\mathrm{m}}}\frac{\tau_{\mathrm{i}}}{r_{\mathrm{f}}}(2l_y-\xi-x),\ x\in[l_y,l_{\mathrm{d}}]\\ \sigma_{\mathrm{mo}}-\left[\sigma_{\mathrm{mo}}-2\frac{V_{\mathrm{f}}}{V_{\mathrm{m}}}\frac{\tau_{\mathrm{f}}}{r_{\mathrm{f}}}(2l_z-\xi)\right.\\ \left.\quad+2\frac{V_{\mathrm{f}}}{V_{\mathrm{m}}}\frac{\tau_{\mathrm{i}}}{r_{\mathrm{f}}}(2l_y-\xi-l_{\mathrm{d}})\right]\exp\left(-\rho\frac{x-l_{\mathrm{d}}}{r_{\mathrm{f}}}\right),\ x\in\left[l_{\mathrm{d}},\frac{l_{\mathrm{c}}}{2}\right]\end{cases}\tag{2.10}$$

The interface shear stress distribution upon reloading is,

$$\tau_{\mathrm{i}}(x)=\begin{cases}\tau_{\mathrm{f}},\ x\in[0,l_z]\\ -\tau_{\mathrm{f}},\ x\in[l_z,\xi]\\ -\tau_{\mathrm{i}},\ x\in[\xi,l_y]\\ \tau_{\mathrm{i}},\ x\in[l_y,l_{\mathrm{d}}]\\ \frac{\rho}{2}\left[\frac{V_{\mathrm{m}}}{V_{\mathrm{f}}}\sigma_{\mathrm{mo}}-\frac{2\tau_{\mathrm{f}}}{r_{\mathrm{f}}}(2l_z-\xi)+\frac{2\tau_{\mathrm{i}}}{r_{\mathrm{f}}}(2l_y-\xi-l_{\mathrm{d}})\right]\exp\left(-\rho\frac{x-l_{\mathrm{d}}}{r_{\mathrm{f}}}\right),\ x\in\left[l_{\mathrm{d}},\frac{l_{\mathrm{c}}}{2}\right]\end{cases}\tag{2.11}$$

2.2.2 *Interface Slip Lengths*

When matrix cracking propagates to the fiber/matrix interface, it deflects along the interface. There are two approaches for the problem of interface debonding, i.e., the interface shear strength approach (Solti et al. 1995) and fracture mechanics approach (Gao et al. 1988). Sun and Singh (Sun and Singh 1998) investigated matrix multiple cracking and fiber/matrix interface debonding by directly observing and analyzing in a transparent SiC fiber-reinforced borosilicate glass composite. For a very short interface debonding length, the shear strength approach and the fracture mechanics approach gave an estimate close to experimental data. However, for a long interface debonding length, the fracture mechanics approach provided a better fit. The fracture mechanics interface debonding criterion is used to determine fiber/matrix interface debonding (Gao et al. 1988).

$$\Gamma_{\rm d} = -\frac{F}{4\pi r_{\rm f}}\frac{\partial w_{\rm f}(0)}{\partial l_{\rm d}} - \frac{1}{2}\int_{0}^{l_{\rm d}} \tau_{\rm i}(x)\frac{\partial v(x)}{\partial l_{\rm d}}{\rm d}x \tag{2.12}$$

where $\Gamma_{\rm d}$ denotes the interface debonding energy, $F(=\pi r_{\rm f}^2\sigma/V_{\rm f})$ denotes the fiber load at the matrix crack plane, $w_{\rm f}(0)$ denotes the fiber axial displacement at the matrix crack plane, and $v(x)$ denotes the relative displacement between the fiber and the matrix.

The fiber and matrix axial displacements $w_{\rm f}(x)$ and $w_{\rm m}(x)$ can be expressed in terms of the fiber and matrix axial stresses:

$$\frac{{\rm d}w_{\rm f}(x)}{{\rm d}x} = \frac{\sigma_{\rm f}(x)}{E_{\rm f}} \tag{2.13}$$

$$\frac{{\rm d}w_{\rm m}(x)}{{\rm d}x} = \frac{\sigma_{\rm m}(x)}{E_{\rm f}} \tag{2.14}$$

The fiber and matrix axial displacements $w_{\rm f}(x)$ and $w_{\rm m}(x)$ are obtained by integrating Eqs. (2.13) and (2.14) from x to $l_{\rm c}/2$, where the fiber and matrix axial stresses from x to $l_{\rm c}/2$ are given by Eqs. (2.1) and (2.2):

$$\begin{aligned} w_{\rm f}(x) &= \int_{x}^{l_{\rm c}/2} \frac{\sigma_{\rm f}(x)}{E_{\rm f}}{\rm d}x \\ &= \frac{\sigma}{V_{\rm f}E_{\rm f}}(l_{\rm d}-x) - \frac{\tau_{\rm f}}{r_{\rm f}E_{\rm f}}\left(2\xi l_{\rm d}-\xi^2-x^2\right) - \frac{\tau_{\rm i}}{r_{\rm f}E_{\rm f}}(l_{\rm d}-\xi)^2 + \frac{\sigma_{\rm fo}}{E_{\rm f}}\left(\frac{l_{\rm c}}{2}-l_{\rm d}\right) \\ &\quad + \frac{r_{\rm f}}{\rho E_{\rm f}}\left[\frac{V_{\rm m}}{V_{\rm f}}\sigma_{\rm mo} - \frac{2\tau_{\rm f}}{r_{\rm f}}\xi - \frac{2\tau_{\rm i}}{r_{\rm f}}(l_{\rm d}-\xi)\right]\left[1-\exp\left(-\rho\frac{l_{\rm c}/2-l_{\rm d}}{r_{\rm f}}\right)\right] \end{aligned} \tag{2.15}$$

$$\begin{aligned} w_{\rm m}(x) &= \int_{x}^{l_{\rm c}} \frac{\sigma_{\rm m}(x)}{E_{\rm m}}{\rm d}x \\ &= \frac{V_{\rm f}\tau_{\rm f}}{r_{\rm f}V_{\rm m}E_m}\left(2\xi l_{\rm d}-\xi^2-x^2\right) + \frac{V_{\rm f}\tau_{\rm i}}{r_{\rm f}V_{\rm m}E_m}(l_{\rm d}-\xi)^2 + \frac{\sigma_{\rm mo}}{E_{\rm m}}\left(\frac{l_{\rm c}}{2}-l_{\rm d}\right) \\ &\quad - \frac{r_{\rm f}}{\rho E_{\rm m}}\left[\sigma_{\rm mo} - 2\frac{V_{\rm f}\tau_{\rm f}}{r_{\rm f}V_{\rm m}}\xi - 2\frac{V_{\rm f}\tau_{\rm i}}{r_{\rm f}V_{\rm m}}(l_{\rm d}-\xi)\right]\left[1-\exp\left(-\rho\frac{l_{\rm c}/2-l_{\rm d}}{r_{\rm f}}\right)\right] \end{aligned} \tag{2.16}$$

The relative displacement $v(x)$ between the fiber and the matrix is

$$
\begin{aligned}
v(x) &= |w_{\mathrm{f}}(x) - w_{\mathrm{m}}(x)| \\
&= \frac{\sigma}{V_{\mathrm{f}}E_{\mathrm{f}}}(l_{\mathrm{d}} - x) - \frac{E_{\mathrm{c}}\tau_{\mathrm{f}}}{r_{\mathrm{f}}V_{\mathrm{m}}E_{\mathrm{m}}E_{\mathrm{f}}}\left(2\xi l_{\mathrm{d}} - \xi^2 - x^2\right) - \frac{E_{\mathrm{c}}\tau_{\mathrm{i}}}{r_{\mathrm{f}}V_{\mathrm{m}}E_{\mathrm{m}}E_{\mathrm{f}}}(l_{\mathrm{d}} - \xi)^2 \\
&+ \frac{r_{\mathrm{f}}E_{\mathrm{c}}}{\rho V_{\mathrm{m}}E_{\mathrm{m}}E_{\mathrm{f}}}\left[\sigma_{\mathrm{mo}} - 2\frac{\tau_{\mathrm{f}}}{r_{\mathrm{f}}}\xi - 2\frac{\tau_{\mathrm{i}}}{r_{\mathrm{f}}}(l_{\mathrm{d}} - \xi)\right]\left[1 - \exp\left(-\rho\frac{l_{\mathrm{c}}/2 - l_{\mathrm{d}}}{r_{\mathrm{f}}}\right)\right]
\end{aligned}
\tag{2.17}
$$

Substituting $w_{\mathrm{f}}(x = 0)$ and $v(x)$ into Eq. (2.14), it leads to the form of

$$
\begin{aligned}
&\frac{E_{\mathrm{c}}\tau_{\mathrm{i}}^2}{r_{\mathrm{f}}V_{\mathrm{m}}E_{\mathrm{m}}E_{\mathrm{f}}}(l_{\mathrm{d}} - \xi)^2 + \frac{E_{\mathrm{c}}\tau_{\mathrm{i}}^2}{\rho V_{\mathrm{m}}E_{\mathrm{m}}E_{\mathrm{f}}}(l_{\mathrm{d}} - \xi) - \frac{\tau_{\mathrm{i}}\sigma}{V_{\mathrm{f}}E_{\mathrm{f}}}(l_{\mathrm{d}} - \xi) + \frac{2E_{\mathrm{c}}\tau_{\mathrm{f}}\tau_{\mathrm{i}}}{r_{\mathrm{f}}V_{\mathrm{m}}E_{\mathrm{m}}E_{\mathrm{f}}}\xi(l_{\mathrm{d}} - \xi) \\
&- \frac{r_{\mathrm{f}}\tau_{\mathrm{i}}\sigma}{2\rho V_{\mathrm{f}}E_{\mathrm{f}}} + \frac{E_{\mathrm{c}}\tau_{\mathrm{f}}^2}{r_{\mathrm{f}}V_{\mathrm{m}}E_{\mathrm{m}}E_{\mathrm{f}}}\xi^2 + \frac{E_{\mathrm{c}}\tau_{\mathrm{f}}\tau_{\mathrm{i}}}{\rho V_{\mathrm{m}}E_{\mathrm{m}}E_{\mathrm{f}}}\xi - \frac{\tau_{\mathrm{f}}\sigma}{V_{\mathrm{f}}E_{\mathrm{f}}}\xi + \frac{r_{\mathrm{f}}V_{\mathrm{m}}E_{\mathrm{m}}\sigma^2}{4V_{\mathrm{f}}^2E_{\mathrm{f}}E_{\mathrm{c}}} - \Gamma_{\mathrm{d}} = 0
\end{aligned}
\tag{2.18}
$$

Solving Eq. (2.18), the interface debonding length l_{d} is

$$
l_{\mathrm{d}} = \left(1 - \frac{\tau_{\mathrm{f}}}{\tau_{\mathrm{i}}}\right)\xi + \frac{r_{\mathrm{f}}}{2}\left(\frac{V_{\mathrm{m}}E_{\mathrm{m}}\sigma}{V_{\mathrm{f}}E_{\mathrm{c}}\tau_{\mathrm{i}}} - \frac{1}{\rho}\right) - \sqrt{\left(\frac{r_{\mathrm{f}}}{2\rho}\right)^2 + \frac{r_{\mathrm{f}}V_{\mathrm{m}}E_{\mathrm{m}}E_{\mathrm{f}}}{E_{\mathrm{c}}\tau_{\mathrm{i}}^2}\Gamma_{\mathrm{d}}}
\tag{2.19}
$$

Upon unloading, the fiber and matrix displacements $w_{\mathrm{f}}(x)$ and $w_{\mathrm{m}}(x)$ are obtained by integrating Eqs. (2.13) and (2.14) from x to $l_{\mathrm{c}}/2$, where the fiber and matrix axial stresses from x to $l_{\mathrm{c}}/2$ are given by Eqs. (2.6) and (2.7):

$$
\begin{aligned}
w_{\mathrm{f}}(x) &= \int_x^{l_{\mathrm{c}}} \frac{\sigma_{\mathrm{f}}(x)}{E_{\mathrm{f}}}\mathrm{d}x = \frac{\sigma}{V_{\mathrm{f}}E_{\mathrm{f}}}(l_{\mathrm{d}} - x) - \frac{\tau_{\mathrm{f}}}{r_{\mathrm{f}}E_{\mathrm{f}}}\left(\xi^2 + x^2 - 2\xi l_y\right) + \frac{2\tau_{\mathrm{f}}}{r_{\mathrm{f}}E_{\mathrm{f}}}\xi(l_{\mathrm{d}} - l_y) \\
&+ \frac{2\tau_{\mathrm{i}}}{r_{\mathrm{f}}E_{\mathrm{f}}}(l_y - \xi)^2 - \frac{\tau_{\mathrm{i}}}{r_{\mathrm{f}}E_{\mathrm{f}}}(2l_y - \xi - l_{\mathrm{d}})^2 + \frac{\sigma_{\mathrm{fo}}}{E_{\mathrm{f}}}\left(\frac{l_{\mathrm{c}}}{2} - l_{\mathrm{d}}\right) \\
&+ \frac{r_{\mathrm{f}}}{\rho E_{\mathrm{f}}}\left[\frac{V_{\mathrm{m}}}{V_{\mathrm{f}}}\sigma_{\mathrm{mo}} + \frac{2\tau_{\mathrm{f}}}{r_{\mathrm{f}}}\xi + \frac{2\tau_{\mathrm{i}}}{r_{\mathrm{f}}}(2l_y - \xi - l_{\mathrm{d}})\right]\left[1 - \exp\left(-\rho\frac{l_{\mathrm{c}}/2 - l_{\mathrm{d}}}{r_{\mathrm{f}}}\right)\right]
\end{aligned}
\tag{2.20}
$$

$$
\begin{aligned}
w_{\mathrm{m}}(x) &= \int_x^{L_{\mathrm{d}}} \frac{\sigma_{\mathrm{m}}(x)}{E_{\mathrm{m}}}\mathrm{d}x = -\frac{V_{\mathrm{f}}\tau_{\mathrm{f}}}{r_{\mathrm{f}}V_{\mathrm{m}}E_m}\left(\xi^2 + x^2 - 2\xi l_{\mathrm{d}}\right) - \frac{2V_{\mathrm{f}}\tau_{\mathrm{i}}}{r_{\mathrm{f}}V_{\mathrm{m}}E_m}(l_y - \xi)^2 \\
&+ \frac{V_{\mathrm{f}}\tau_{\mathrm{i}}}{r_{\mathrm{f}}V_{\mathrm{m}}E_m}(2l_y - \xi - l_{\mathrm{d}})^2 + \frac{\sigma_{\mathrm{mo}}}{E_{\mathrm{m}}}\left(\frac{l_{\mathrm{c}}}{2} - l_{\mathrm{d}}\right) \\
&- \frac{r_{\mathrm{f}}}{\rho E_{\mathrm{m}}}\left[\sigma_{\mathrm{mo}} + 2\frac{V_{\mathrm{f}}}{V_{\mathrm{m}}}\frac{\tau_{\mathrm{f}}}{r_{\mathrm{f}}}\xi + 2\frac{V_{\mathrm{f}}}{V_{\mathrm{m}}}\frac{\tau_{\mathrm{i}}}{r_{\mathrm{f}}}(2l_y - \xi - l_{\mathrm{d}})\right]\left[1 - \exp\left(-\rho\frac{l_{\mathrm{c}}/2 - l_{\mathrm{d}}}{r_{\mathrm{f}}}\right)\right]
\end{aligned}
\tag{2.21}
$$

The relative displacement $v(x)$ between the fiber and the matrix is

$$\begin{aligned} v(x) = |w_{\rm f}(x) - w_{\rm m}(x)| &= \frac{\sigma}{V_{\rm f}E_{\rm f}}(l_{\rm d} - x) - \frac{\tau_{\rm f}E_{\rm c}}{r_{\rm f}V_{\rm m}E_{\rm m}E_{\rm f}}\left(x^2 + \xi^2 - 2\xi l_{\rm d}\right) \\ &+ \frac{2E_{\rm c}\tau_{\rm i}}{r_{\rm f}V_{\rm m}E_{\rm m}E_{\rm f}}\left(l_y - \xi\right)^2 - \frac{\tau_{\rm i}E_{\rm c}}{r_{\rm f}V_{\rm m}E_{\rm m}E_{\rm f}}\left(2l_y - \xi - l_{\rm d}\right)^2 \\ &+ \frac{r_{\rm f}}{\rho E_{\rm f}}\left[\frac{V_{\rm m}}{V_{\rm f}}\sigma_{\rm mo} + \frac{2\tau_{\rm f}}{r_{\rm f}}\xi + \frac{2\tau_{\rm i}}{r_{\rm f}}\left(2l_y - \xi - l_{\rm d}\right)\right]\left[1 - \exp\left(-\rho\frac{l_{\rm c}/2 - l_{\rm d}}{r_{\rm f}}\right)\right] \\ &+ \frac{r_{\rm f}}{\rho E_{\rm m}}\left[\sigma_{\rm mo} + 2\frac{V_{\rm f}}{V_{\rm m}}\frac{\tau_{\rm f}}{r_{\rm f}}\xi + 2\frac{V_{\rm f}}{V_{\rm m}}\frac{\tau_{\rm i}}{r_{\rm f}}\left(2l_y - \xi - l_{\rm d}\right)\right]\left[1 - \exp\left(-\rho\frac{l_{\rm c}/2 - l_{\rm d}}{r_{\rm f}}\right)\right] \end{aligned} \tag{2.22}$$

Substituting $w_{\rm f}(x = 0)$ and $v(x)$ into Eq. (2.12), it leads to the form of

$$\begin{aligned} &\frac{r_{\rm f}V_{\rm m}E_{\rm m}}{4V_{\rm f}^2E_{\rm f}E_{\rm c}}\sigma^2 + \frac{\tau_{\rm f}\sigma}{V_{\rm f}E_{\rm f}}\xi + \frac{\tau_{\rm i}\sigma}{V_{\rm f}E_{\rm f}}\left(2l_y - \xi - l_{\rm d}\right) - \frac{r_{\rm f}\tau_{\rm i}\sigma}{2\rho V_{\rm f}E_{\rm f}} \\ &+ \frac{E_{\rm c}\tau_{\rm f}^2}{r_{\rm f}V_{\rm m}E_{\rm m}E_{\rm f}}\xi^2 + \frac{2E_{\rm c}\tau_{\rm f}\tau_{\rm i}}{r_{\rm f}V_{\rm m}E_{\rm m}E_{\rm f}}\xi\left(2l_y - \xi - l_{\rm d}\right) - \frac{E_{\rm c}\tau_{\rm f}\tau_{\rm i}}{\rho V_{\rm m}E_{\rm m}E_{\rm f}}\xi \\ &+ \frac{E_{\rm c}\tau_{\rm i}^2}{r_{\rm f}V_{\rm m}E_{\rm m}E_{\rm f}}\left(2l_y - \xi - l_{\rm d}\right)^2 - \frac{E_{\rm c}\tau_{\rm i}^2}{\rho V_{\rm m}E_{\rm m}E_{\rm f}}\left(2l_y - \xi - l_{\rm d}\right) - \Gamma_{\rm d} = 0 \end{aligned} \tag{2.23}$$

Solving Eq. (2.23), the unloading interface counter-slip length l_y is

$$\begin{aligned} l_y = \frac{1}{2}\Bigg\{ l_{\rm d}(\sigma_{\max}) + \left(1 - \frac{\tau_{\rm f}}{\tau_{\rm i}}\right)\xi - \Bigg[\frac{r_{\rm f}}{2}\left(\frac{V_{\rm m}E_{\rm m}}{V_{\rm f}E_{\rm c}}\frac{\sigma}{\tau_{\rm i}} - \frac{1}{\rho}\right) \\ - \sqrt{\left(\frac{r_{\rm f}}{2\rho}\right)^2 + \frac{r_{\rm f}V_{\rm m}E_{\rm m}E_{\rm f}}{E_{\rm c}\tau_{\rm i}^2}\Gamma_{\rm d}}\Bigg]\Bigg\} \end{aligned} \tag{2.24}$$

Upon reloading, the fiber and matrix displacements $w_{\rm f}(x)$ and $w_{\rm m}(x)$ are obtained by integrating Eqs. (2.13) and (2.14) from x to $l_{\rm c}/2$, where the fiber and matrix axial stresses from x to $l_{\rm c}/2$ are given by Eq. (2.9) and (2.10):

$$\begin{aligned} w_{\rm f}(x) = \int_x^{l_{\rm d}}\frac{\sigma_{\rm f}(x)}{E_{\rm f}}{\rm d}x &= \frac{\sigma}{V_{\rm f}E_{\rm f}}(l_{\rm d} - x) - \frac{\tau_{\rm f}}{r_{\rm f}E_{\rm f}}\left(2l_z^2 - x^2\right) + \frac{\tau_{\rm f}}{r_{\rm f}E_{\rm f}}(2l_z - \xi)^2 \\ &- \frac{2\tau_{\rm f}}{r_{\rm f}E_{\rm f}}(2l_z - \xi)(l_{\rm d} - \xi) + \frac{2\tau_{\rm i}}{r_{\rm f}E_{\rm f}}\left(l_y - \xi\right)^2 - \frac{\tau_{\rm i}}{r_{\rm f}E_{\rm f}}\left(2l_y - \xi - l_{\rm d}\right)^2 \\ &+ \frac{\sigma_{\rm fo}}{E_{\rm f}}\left(\frac{l_{\rm c}}{2} - l_{\rm d}\right) + \frac{r_{\rm f}}{\rho E_{\rm f}}\left[\frac{V_{\rm m}}{V_{\rm f}}\sigma_{\rm mo} - \frac{2\tau_{\rm f}}{r_{\rm f}}(2l_z - \xi)\right. \\ &\left.+ \frac{2\tau_{\rm i}}{r_{\rm f}}\left(2l_y - \xi - l_{\rm d}\right)\right]\left[1 - \exp\left(-\rho\frac{l_{\rm c}/2 - l_{\rm d}}{r_{\rm f}}\right)\right] \end{aligned} \tag{2.25}$$

$$
\begin{aligned}
w_{\mathrm{m}}(x) &= \int_{x}^{l_{\mathrm{d}}} \frac{\sigma_{\mathrm{m}}(x)}{E_{\mathrm{m}}} \mathrm{d}x \\
&= \frac{V_{\mathrm{f}}\tau_{\mathrm{f}}}{r_{\mathrm{f}}V_{\mathrm{m}}E_{m}}\left(2l_{z}^{2} - x^{2}\right) - \frac{V_{\mathrm{f}}\tau_{\mathrm{f}}}{r_{\mathrm{f}}V_{\mathrm{m}}E_{m}}(2l_{z} - \xi)^{2} \\
&+ \frac{2V_{\mathrm{f}}\tau_{\mathrm{f}}}{r_{\mathrm{f}}V_{\mathrm{m}}E_{m}}(2l_{z} - \xi)(l_{\mathrm{d}} - \xi) - \frac{2V_{\mathrm{f}}\tau_{\mathrm{i}}}{r_{\mathrm{f}}V_{\mathrm{m}}E_{m}}\left(l_{y} - \xi\right)^{2} \\
&+ \frac{V_{\mathrm{f}}\tau_{\mathrm{i}}}{r_{\mathrm{f}}V_{\mathrm{m}}E_{m}}\left(2l_{y} - \xi - l_{\mathrm{d}}\right)^{2} + \frac{\sigma_{\mathrm{mo}}}{E_{\mathrm{m}}}\left(\frac{l_{\mathrm{c}}}{2} - l_{\mathrm{d}}\right) \\
&- \frac{r_{\mathrm{f}}}{\rho E_{\mathrm{m}}}\left[\sigma_{\mathrm{mo}} - 2\frac{V_{\mathrm{f}}}{V_{\mathrm{m}}}\frac{\tau_{\mathrm{f}}}{r_{\mathrm{f}}}(2l_{z} - \xi)\right. \\
&\left. + 2\frac{V_{\mathrm{f}}}{V_{\mathrm{m}}}\frac{\tau_{\mathrm{i}}}{r_{\mathrm{f}}}\left(2l_{y} - \xi - l_{\mathrm{d}}\right)\right]\left[1 - \exp\left(-\rho\frac{l_{\mathrm{c}}/2 - l_{\mathrm{d}}}{r_{\mathrm{f}}}\right)\right]
\end{aligned} \tag{2.26}
$$

The relative displacement $v(x)$ between the fiber and the matrix is

$$
\begin{aligned}
v(x) &= |w_{\mathrm{f}}(x) - w_{\mathrm{m}}(x)| \\
&= \frac{\sigma}{V_{\mathrm{f}}E_{\mathrm{f}}}(l_{\mathrm{d}} - x) - \frac{E_{\mathrm{c}}\tau_{\mathrm{f}}}{r_{\mathrm{f}}V_{\mathrm{m}}E_{\mathrm{m}}E_{\mathrm{f}}}\left(2l_{z}^{2} - x^{2}\right) + \frac{E_{\mathrm{c}}\tau_{\mathrm{f}}}{r_{\mathrm{f}}V_{\mathrm{m}}E_{\mathrm{m}}E_{\mathrm{f}}}(2l_{z} - \xi)^{2} \\
&- \frac{2E_{\mathrm{c}}\tau_{\mathrm{f}}}{r_{\mathrm{f}}V_{\mathrm{m}}E_{\mathrm{m}}E_{\mathrm{f}}}(2l_{z} - \xi)(l_{\mathrm{d}} - \xi) + \frac{2E_{\mathrm{c}}\tau_{\mathrm{i}}}{r_{\mathrm{f}}V_{\mathrm{m}}E_{\mathrm{m}}E_{\mathrm{f}}}\left(l_{y} - \xi\right)^{2} \\
&- \frac{E_{\mathrm{c}}\tau_{\mathrm{i}}}{r_{\mathrm{f}}V_{\mathrm{m}}E_{\mathrm{m}}E_{\mathrm{f}}}\left(2l_{y} - \xi - l_{\mathrm{d}}\right)^{2} \\
&+ \frac{r_{\mathrm{f}}}{\rho E_{\mathrm{f}}}\left[\frac{V_{\mathrm{m}}}{V_{\mathrm{f}}}\sigma_{\mathrm{mo}} - \frac{2\tau_{\mathrm{f}}}{r_{\mathrm{f}}}(2l_{z} - \xi)\right. \\
&\left. + \frac{2\tau_{\mathrm{i}}}{r_{\mathrm{f}}}\left(2l_{y} - \xi - l_{\mathrm{d}}\right)\right]\left[1 - \exp\left(-\rho\frac{l_{\mathrm{c}}/2 - l_{\mathrm{d}}}{r_{\mathrm{f}}}\right)\right] \\
&+ \frac{r_{\mathrm{f}}}{\rho E_{\mathrm{m}}}\left[\sigma_{\mathrm{mo}} - 2\frac{V_{\mathrm{f}}}{V_{\mathrm{m}}}\frac{\tau_{\mathrm{f}}}{r_{\mathrm{f}}}(2l_{z} - \xi)\right. \\
&\left. + 2\frac{V_{\mathrm{f}}}{V_{\mathrm{m}}}\frac{\tau_{\mathrm{i}}}{r_{\mathrm{f}}}\left(2l_{y} - \xi - l_{\mathrm{d}}\right)\right]\left[1 - \exp\left(-\rho\frac{l_{\mathrm{c}}/2 - l_{\mathrm{d}}}{r_{\mathrm{f}}}\right)\right]
\end{aligned} \tag{2.27}
$$

Substituting $w_{\mathrm{f}}(x = 0)$ and $v(x)$ into Eq. (2.12), it leads to the form of

$$
\begin{aligned}
&\frac{r_{\mathrm{f}}V_{\mathrm{m}}E_{\mathrm{m}}\sigma^{2}}{4V_{\mathrm{f}}^{2}E_{\mathrm{f}}E_{\mathrm{c}}} - \frac{\tau_{\mathrm{f}}\sigma}{V_{\mathrm{f}}E_{\mathrm{f}}}(2l_{z} - \xi) + \frac{\tau_{\mathrm{i}}\sigma}{V_{\mathrm{f}}E_{\mathrm{f}}}\left(2l_{y} - \xi - l_{\mathrm{d}}\right) - \frac{r_{\mathrm{f}}\tau_{\mathrm{i}}\sigma}{2\rho V_{\mathrm{f}}E_{\mathrm{f}}} \\
&+ \frac{E_{\mathrm{c}}\tau_{\mathrm{f}}^{2}}{r_{\mathrm{f}}V_{\mathrm{m}}E_{\mathrm{m}}E_{\mathrm{f}}}(2l_{z} - \xi)^{2} - \frac{2E_{\mathrm{c}}\tau_{\mathrm{f}}\tau_{\mathrm{i}}}{r_{\mathrm{f}}V_{\mathrm{m}}E_{\mathrm{m}}E_{\mathrm{f}}}(2l_{z} - \xi)\left(2l_{y} - \xi - l_{\mathrm{d}}\right) \\
&+ \frac{E_{\mathrm{c}}\tau_{\mathrm{f}}\tau_{\mathrm{i}}}{\rho V_{\mathrm{m}}E_{\mathrm{m}}E_{\mathrm{f}}}(2l_{z} - \xi) + \frac{E_{\mathrm{c}}\tau_{\mathrm{i}}^{2}}{r_{\mathrm{f}}V_{\mathrm{m}}E_{\mathrm{m}}E_{\mathrm{f}}}\left(2l_{y} - \xi - l_{\mathrm{d}}\right)^{2}
\end{aligned}
$$

$$-\frac{E_{\mathrm{c}}\tau_{\mathrm{i}}^{2}}{\rho V_{\mathrm{m}}E_{\mathrm{m}}E_{\mathrm{f}}}(2l_{y}-\xi-l_{\mathrm{d}})-\Gamma_{\mathrm{d}}=0 \tag{2.28}$$

Solving Eq. (2.28), the reloading interface new-slip length l_z is

$$\begin{aligned} l_{z}=\frac{\tau_{\mathrm{i}}}{\tau_{\mathrm{f}}}\Bigg\{l_{y}(\sigma_{\min})-\frac{1}{2}\Bigg[l_{\mathrm{d}}+\left(1-\frac{\tau_{\mathrm{f}}}{\tau_{\mathrm{i}}}\right)\xi-\Bigg[\frac{r_{\mathrm{f}}}{2}\left(\frac{V_{\mathrm{m}}E_{\mathrm{m}}}{V_{\mathrm{f}}E_{\mathrm{c}}}\frac{\sigma}{\tau_{\mathrm{i}}}-\frac{1}{\rho}\right) \\ -\sqrt{\left(\frac{r_{\mathrm{f}}}{2\rho}\right)^{2}+\frac{r_{\mathrm{f}}V_{\mathrm{m}}E_{\mathrm{m}}E_{\mathrm{f}}}{E_{\mathrm{c}}\tau_{\mathrm{i}}^{2}}\Gamma_{\mathrm{d}}}\Bigg]\Bigg]\Bigg\} \end{aligned} \tag{2.29}$$

2.2.3 Hysteresis Constitutive Relationship

When damage forms within the composite, the composite strain is determined by Eq. (2.30), which assumes that the composite strain is equivalent to the average strain in an undamaged fiber.

$$\varepsilon_{\mathrm{c}}=\frac{2}{E_{\mathrm{f}}l_{\mathrm{c}}}\int_{l_{\mathrm{c}}/2}\sigma_{\mathrm{f}}(x)\mathrm{d}x-(\alpha_{\mathrm{c}}-\alpha_{\mathrm{f}})\Delta T \tag{2.30}$$

The unloading and reloading stress−strain relationships are obtained by integrating Eq. (2.30) from zero to $l_c/2$, where the fiber axial stresses from zero to $l_c/2$ upon unloading and reloading are given by Eqs. (2.6) and (2.9):

$$\begin{aligned} \varepsilon_{\mathrm{c\_unload}}= & \frac{2\sigma l_{\mathrm{d}}}{V_{\mathrm{f}}E_{\mathrm{f}}l_{\mathrm{c}}}+\frac{2\tau_{\mathrm{f}}}{r_{\mathrm{f}}E_{\mathrm{f}}l_{\mathrm{c}}}\xi^{2}+\frac{4\tau_{\mathrm{f}}}{r_{\mathrm{f}}E_{\mathrm{f}}l_{\mathrm{c}}}\xi(l_{\mathrm{d}}-\xi)+\frac{4\tau_{\mathrm{i}}}{r_{\mathrm{f}}E_{\mathrm{f}}l_{\mathrm{c}}}(l_{y}-\xi)^{2} \\ & -\frac{2\tau_{\mathrm{i}}}{r_{\mathrm{f}}E_{\mathrm{f}}l_{\mathrm{c}}}(2l_{y}-\xi-l_{\mathrm{d}})^{2}+\frac{2\sigma_{\mathrm{fo}}}{E_{\mathrm{f}}l_{\mathrm{c}}}\left(\frac{l_{\mathrm{c}}}{2}-l_{\mathrm{d}}\right) \\ & +\frac{2r_{\mathrm{f}}}{\rho E_{\mathrm{f}}l_{\mathrm{c}}}\left[\frac{V_{\mathrm{m}}}{V_{\mathrm{f}}}\sigma_{\mathrm{mo}}+\frac{2\tau_{\mathrm{f}}}{r_{\mathrm{f}}}\xi+\frac{2\tau_{\mathrm{i}}}{r_{\mathrm{f}}}(2l_{y}-\xi-l_{\mathrm{d}})\right] \\ & \times\left[1-\exp\left(-\rho\frac{l_{\mathrm{c}}/2-l_{\mathrm{d}}}{r_{\mathrm{f}}}\right)\right]-(\alpha_{c}-\alpha_{f})\Delta\mathrm{T} \end{aligned} \tag{2.31}$$

$$\begin{aligned} \varepsilon_{\mathrm{c\_reload}}= & \frac{2\sigma}{V_{\mathrm{f}}E_{\mathrm{f}}l_{\mathrm{c}}}l_{\mathrm{d}}-\frac{4\tau_{\mathrm{f}}}{r_{\mathrm{f}}E_{\mathrm{f}}l_{\mathrm{c}}}l_{z}^{2}+\frac{2\tau_{\mathrm{f}}}{r_{\mathrm{f}}E_{\mathrm{f}}l_{\mathrm{c}}}(2l_{z}-\xi)^{2}-\frac{4\tau_{\mathrm{f}}}{r_{\mathrm{f}}E_{\mathrm{f}}l_{\mathrm{c}}}(2l_{z}-\xi)(l_{\mathrm{d}}-\xi) \\ & +\frac{4\tau_{\mathrm{i}}}{r_{\mathrm{f}}E_{\mathrm{f}}l_{\mathrm{c}}}(l_{y}-\xi)^{2}-\frac{2\tau_{\mathrm{i}}}{r_{\mathrm{f}}E_{\mathrm{f}}l_{\mathrm{c}}}(2l_{y}-\xi-l_{\mathrm{d}})^{2}+\frac{2\sigma_{\mathrm{fo}}}{E_{\mathrm{f}}l_{\mathrm{c}}}\left(\frac{l_{\mathrm{c}}}{2}-l_{\mathrm{d}}\right) \\ & +\frac{2r_{\mathrm{f}}}{\rho E_{\mathrm{f}}l_{\mathrm{c}}}\left[\frac{V_{\mathrm{m}}}{V_{\mathrm{f}}}\sigma_{\mathrm{mo}}-\frac{2\tau_{\mathrm{f}}}{r_{\mathrm{f}}}(2l_{z}-\xi)+\frac{2\tau_{\mathrm{i}}}{r_{\mathrm{f}}}(2l_{y}-\xi-l_{\mathrm{d}})\right] \end{aligned}$$

$$\times \left[1 - \exp\left(-\rho \frac{l_c/2 - l_d}{r_f} \right) \right] - (\alpha_c - \alpha_f)\Delta T \quad (2.32)$$

2.3 Results and Discussion

The fatigue hysteresis loops of fiber-reinforced CMCs under stochastic loading are related to the material properties, stress levels, and matrix cracking, i.e., fatigue peak stress, fiber volume, matrix crack spacing, interface debonding, interface wear, and stochastic loading sequence. Effects of these factors on fatigue hysteresis loops are analyzed. The ceramic composite system of SiC/CAS is used for the case study and its basic material properties are given by: $V_f = 35\%$, $E_f = 190$ GPa, $E_m = 90$ GPa, $r_f = 7.5$ μm, $\alpha_f = 4 \times 10^{-6}$/°C, $\alpha_m = 5 \times 10^{-6}$/°C, $\Delta T = -1000$ °C, $\tau_i = 15$ MPa, $\tau_f = 5$ MPa, and $\Gamma_d = 0.1$ J/m^2.

2.3.1 *Hysteresis Loops for Different Stochastic Fatigue Peak Stresses*

Figure 2.1 shows the effect of fatigue peak stress, i.e., σ_{max} = 180, 240, and 300 MPa, on fatigue hysteresis loops, in which the unloading residual strain and fatigue hysteresis loops area, i.e., fatigue hysteresis loss energy, increase, and the fatigue hysteresis modulus decreases with the increase of fatigue peak stress.

Figure 2.2 shows the fatigue hysteresis loops and unloading/reloading interface slip between the fiber and the matrix with and without considering interface wear at σ_{max} = 180 MPa, in which the fatigue hysteresis dissipated energy increases from U

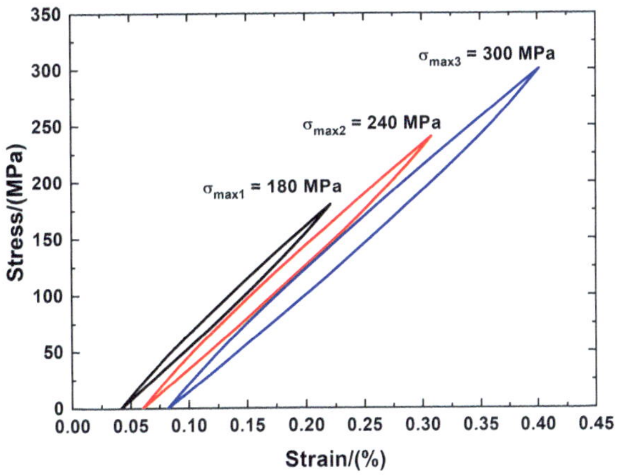

Fig. 2.1 Effect of fatigue peak stress on fatigue hysteresis loops

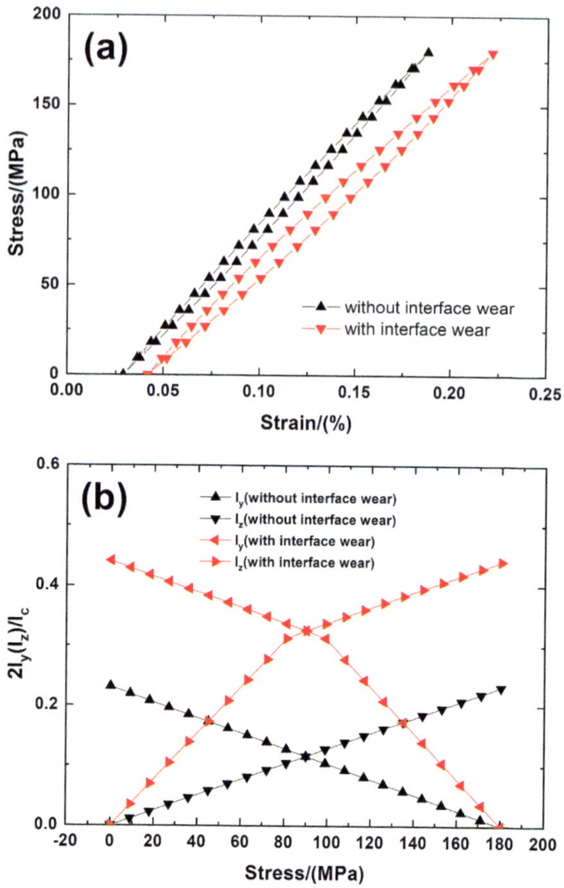

Fig. 2.2 **a** Fatigue hysteresis loops; **b** the interface slip length versus applied stress with and without considering interface wear at σ_{max} = 180 MPa

= 8.7 kJ/m^3 at the first loading to U = 16.3 kJ/m^3 when interface shear stress in the debonding region decreases from initial value τ_i = 15 MPa to τ_f = 5 MPa, as shown in Fig. 2.2a; the unloading interface counter-slip ratio $2l_y/l_c$ and reloading interface new-slip length $2l_z/l_c$ increases from 23% at the first loading to 44% when interface shear stress in the debonding region decreases from initial value τ_i = 15 MPa to τ_f = 5 MPa, as shown in Fig. 2.2b.

Under stochastic fatigue peak stress σ_{max} = 240 MPa, the fatigue hysteresis dissipated energy increases from U = 8.7 kJ/m^3 at σ_{max} = 180 MPa to U = 31.9 kJ/m^3 under σ_{max} = 240 MPa considering interface wear in the interface debonding region, and to U = 20.7 kJ/m^3 under σ_{max} = 240 MPa without considering interface wear, as shown in Fig. 2.3a; the unloading interface counter-slip length $2l_y/l_c$ and reloading interface new-slip length $2l_z/l_c$ increases from 23% at σ_{max} = 180 MPa to 52% under σ_{max} = 240 MPa considering interface wear in the interface debonding region, and to 31% under σ_{max} = 240 MPa without considering interface wear, as shown in Fig. 2.3b.

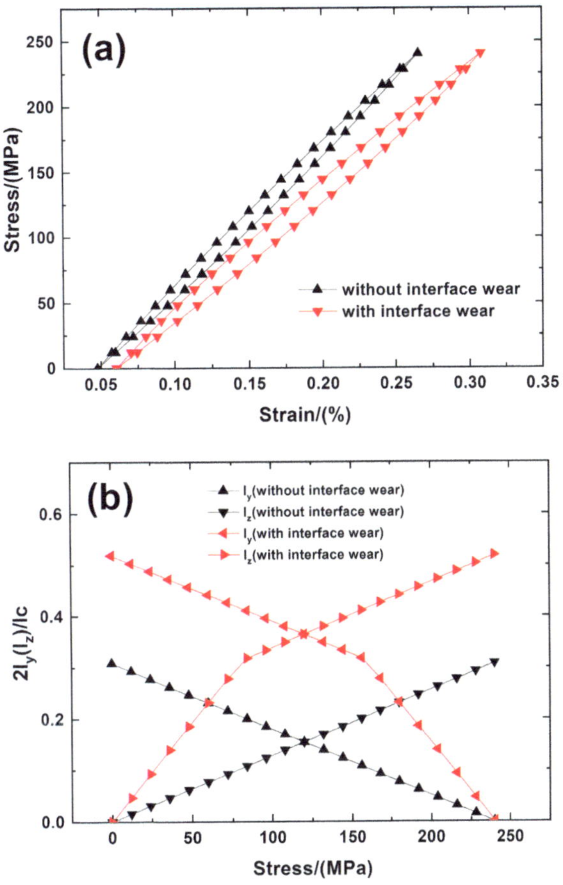

Fig. 2.3 **a** Fatigue hysteresis loops; **b** the interface slip length versus applied stress with and without considering interface wear at $\sigma_{max} =$ 240 MPa

Under stochastic fatigue peak stress $\sigma_{max} = 300$ MPa, the fatigue hysteresis dissipated energy increases from $U = 8.7$ kJ/m^3 at the first loading under $\sigma_{max} = 180$ MPa to $U = 55.3$ kJ/m^3 under $\sigma_{max} = 300$ MPa considering interface wear in the interface debonding region, and to $U = 40.6$ kJ/m^3 under $\sigma_{max} = 300$ MPa without considering interface wear, as shown in Fig. 2.4a; the unloading interface counter-slip length $2l_y/l_c$ and reloading interface new-slip length $2l_z/l_c$ decreases from 23% at the first loading under $\sigma_{max} = 180$ MPa to 60% under $\sigma_{max} = 300$ MPa considering interface wear in the interface debonding region, and to 38% under $\sigma_{max} = 300$ MPa without considering interface wear, as shown in Fig. 2.4b.

When stochastic fatigue peak stress increased, the unloading/reloading interface slip proportion, i.e., $2l_y/l_c$ and $2l_z/l_c$, increased, leading to the increase of unloading residual strain and fatigue hysteresis loops area, and the decrease of fatigue hysteresis modulus.

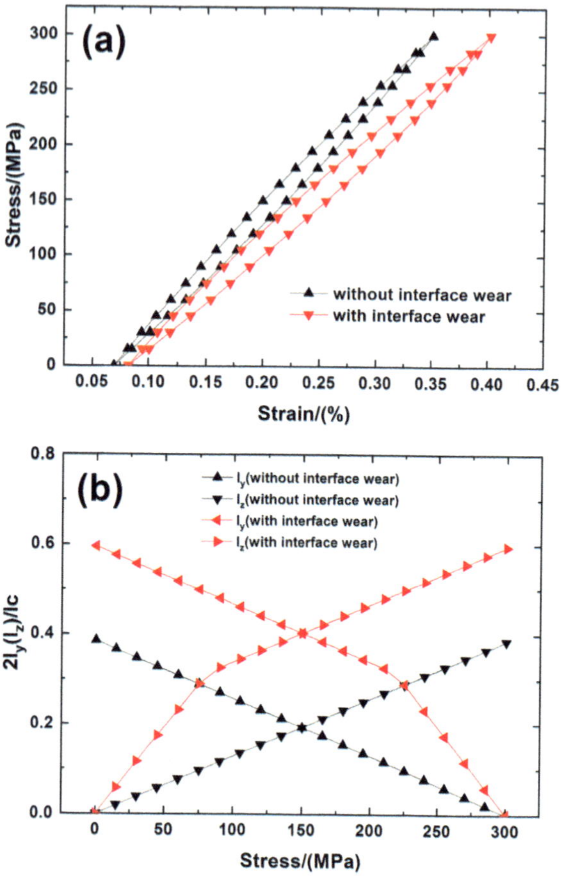

Fig. 2.4 **a** Fatigue hysteresis loops; **b** the interface slip length versus applied stress with and without considering interface wear at $\sigma_{\max} =$ 300 MPa

2.3.2 Hysteresis Loops Under Stochastic Loading for Different Fiber Volumes

Figure 2.5 shows the effect of fiber volume (i.e., $V_f = 35\%$ and 40%) on fatigue hysteresis loops and unloading/reloading interface slip between the fiber and the matrix under constant fatigue peak stress $\sigma_{\max} = 180$ MPa and stochastic fatigue peak stress $\sigma_{\max} = 300$ MPa with interface shear stress $\tau_f = 5$ MPa in the interface wear region. When the fiber volume increases from 35 to 40%, the fatigue hysteresis loops area and unloading residual strain decreases, i.e., the fatigue hysteresis dissipated energy decreases from $U = 55.3$ to 31.9 kJ/m^3, and the fatigue hysteresis modulus increases; the unloading interface counter-slip length $2l_y/l_c$ and reloading interface new-slip length $2l_z/l_c$ decreases from 59.5% to 44.7%.

When the fiber volume fraction increased, the unloading/reloading interface slip proportion, i.e., $2l_y/l_c$ and $2l_z/l_c$, decreased, leading to the decrease of unloading

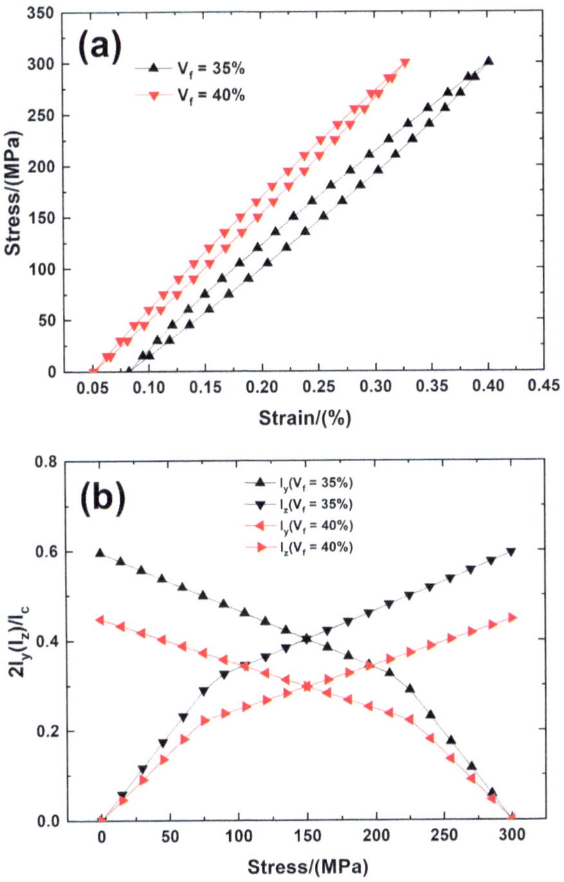

Fig. 2.5 Effect of the fiber volume (i.e., $V_f = 35\%$ and 40%) on **a** the fatigue hysteresis loops; **b** the interface slip ratio versus applied stress

residual strain and fatigue hysteresis loops area, and the increase of fatigue hysteresis modulus.

2.3.3 Hysteresis Loops Under Stochastic Loading for Different Matrix Crack Spacing

Figure 2.6 shows the effect of matrix crack spacing (i.e., $l_c = 30r_f$ and $40r_f$) on fatigue hysteresis loops and unloading/reloading interface slip between the fiber and the matrix under constant fatigue peak stress $\sigma_{max} = 180$ MPa and stochastic fatigue peak stress $\sigma_{max} = 300$ MPa with interface shear stress $\tau_f = 5$ MPa in the interface wear region. When matrix crack spacing increases from $l_c = 30r_f$ to $40r_f$, the fatigue hysteresis loops area and unloading residual strain decrease, i.e., the fatigue hysteresis dissipated energy decreases from $U = 63.9$ to 47.9 kJ/m^3, and the fatigue hysteresis

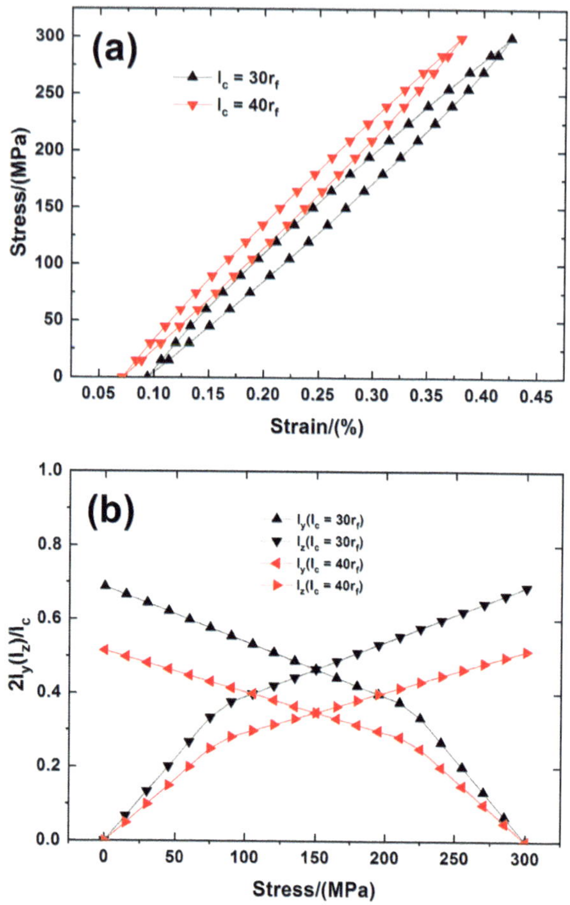

Fig. 2.6 Effect of matrix crack spacing (i.e., $l_c = 30r_f$ and $40r_f$) on **a** the fatigue hysteresis loops; **b** the interface slip ratio versus applied stress

modulus increases; the unloading interface counter-slip length $2l_y/l_c$ and reloading interface new-slip length $2l_z/l_c$ decreases from 68.8% to 51.6%.

When matrix crack spacing increased, the unloading/reloading interface slip proportion, i.e., $2l_y/l_c$ and $2l_z/l_c$, decreased, leading to the decrease of unloading residual strain and fatigue hysteresis loops area, and the increase of fatigue hysteresis modulus.

2.3.4 Hysteresis Loops Under Stochastic Loading for Different Interface Properties

Figure 2.7 shows the effect of interface shear stress (i.e., $\tau_i = 15$ and 20 MPa) on fatigue hysteresis loops and unloading/reloading interface slip between the fiber and the matrix under constant fatigue peak stress $\sigma_{max} = 180$ MPa and stochastic fatigue peak stress $\sigma_{max} = 300$ MPa with interface shear stress $\tau_f = 5$ MPa in the interface wear region. When the interface shear stress increases from $\tau_i = 15$ to 20 MPa, the unloading residual strain and fatigue hysteresis loops area decrease, i.e., the fatigue hysteresis dissipated energy decreases from $U = 55.3$ to 40 kJ/m^3, and the fatigue hysteresis modulus increases; the unloading interface counter-slip length $2l_y/l_c$ and reloading interface new-slip length $2l_z/l_c$ decreases from 59.5% to 46.3%. When the interface shear stress increased, the unloading/reloading interface slip proportion, i.e., $2l_y/l_c$ and $2l_z/l_c$, decreased, leading to the decrease of unloading residual strain and fatigue hysteresis loops area, and the increase of fatigue hysteresis modulus.

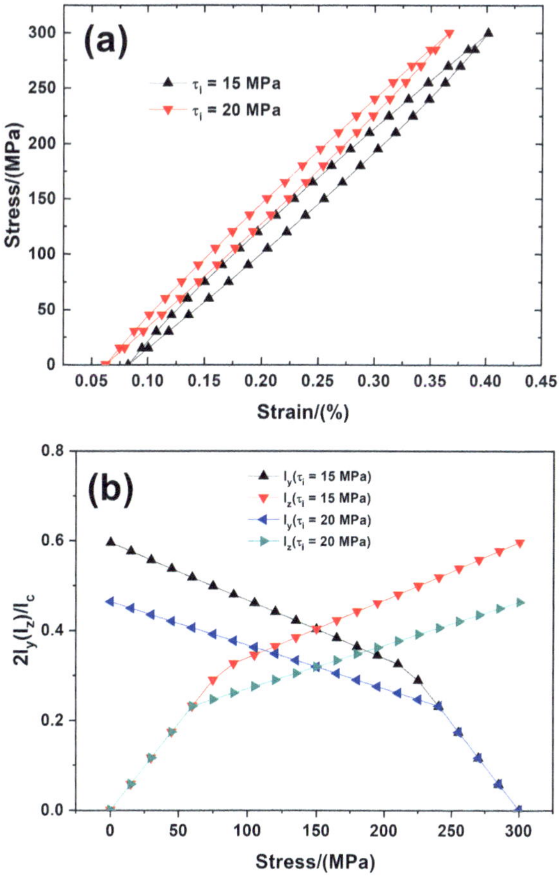

Fig. 2.7 Effect of interface shear stress (i.e., $\tau_i = 15$ and 20 MPa) on **a** the fatigue hysteresis loops; **b** the interface slip ratio versus applied stress

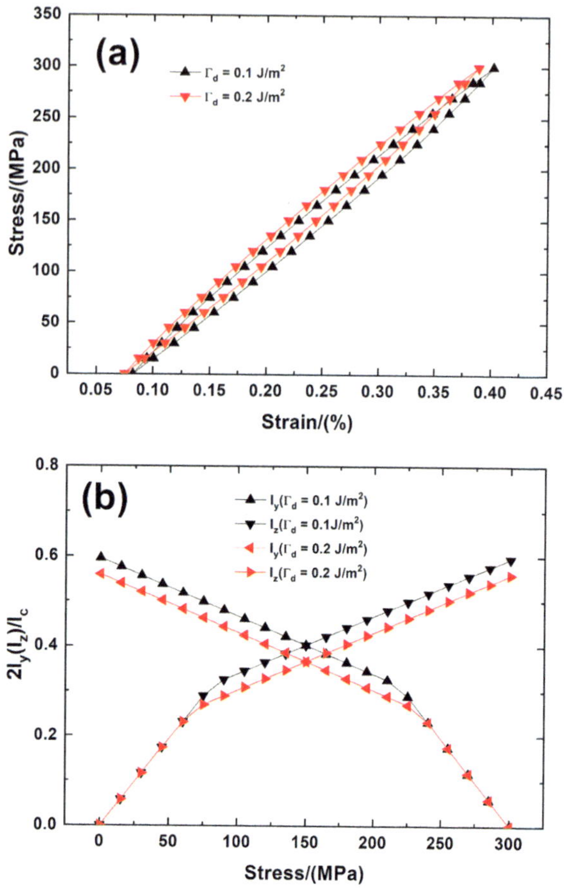

Fig. 2.8 Effect of interface debonding energy (i.e., Γ_d = 0.1 and 0.2 J/m^2) on **a** the fatigue hysteresis loops; **b** the interface slip ratio versus applied stress

Figure 2.8 shows the effect of interface debonding energy (i.e., $\Gamma_d = 0.1$ and 0.2 J/m^2) on fatigue hysteresis loops and unloading/reloading interface slip between the fiber and the matrix under constant fatigue peak stress $\sigma_{max} = 180$ MPa and stochastic fatigue peak stress $\sigma_{max} = 300$ MPa with interface shear stress $\tau_f =$ 5 MPa in the interface wear region. When the interface debonding energy increases from $\Gamma_d = 0.1$ to 0.2 J/m^2, the unloading residual strain and fatigue hysteresis loops area decrease, i.e., the fatigue hysteresis dissipated energy decreases from U = 55.3 to 51 kJ/m^3, and the fatigue hysteresis modulus increases; the unloading interface counter-slip length $2l_y/l_c$ and reloading interface new-slip length $2l_z/l_c$ decreases from 59.5% to 55.9%. When the interface debonding energy increased, the unloading/reloading interface slip proportion, i.e., $2l_y/l_c$ and $2l_z/l_c$, decreased, leading to the decrease of unloading residual strain and fatigue hysteresis loops area, and the increase of fatigue hysteresis modulus.

2.3.5 Hysteresis Loops Under Stochastic Loading Considering Interface Wear

Figure 2.9 shows the effect of interface shear stress in the interface wear region (i.e., $\tau_f = 1$ and 5 MPa) on fatigue hysteresis loops and unloading/reloading interface slip between the fiber and the matrix under constant fatigue peak stress $\sigma_{max} = 180$ MPa and stochastic fatigue peak stress $\sigma_{max} = 300$ MPa with interface shear stress $\tau_i =$ 15 MPa in the interface debonding region. When the interface shear stress in the interface wear region increases from $\tau_f = 5$ to 10 MPa, the fatigue hysteresis loops area and unloading residual strain decrease, i.e., the fatigue hysteresis dissipated energy decreases from $U = 55.3$ to 52.1 kJ/m^3, and the fatigue hysteresis modulus increases; the unloading interface counter-slip length $2l_y/l_c$ and reloading interface new-slip length $2l_z/l_c$ decreases from 59.5% to 49%.

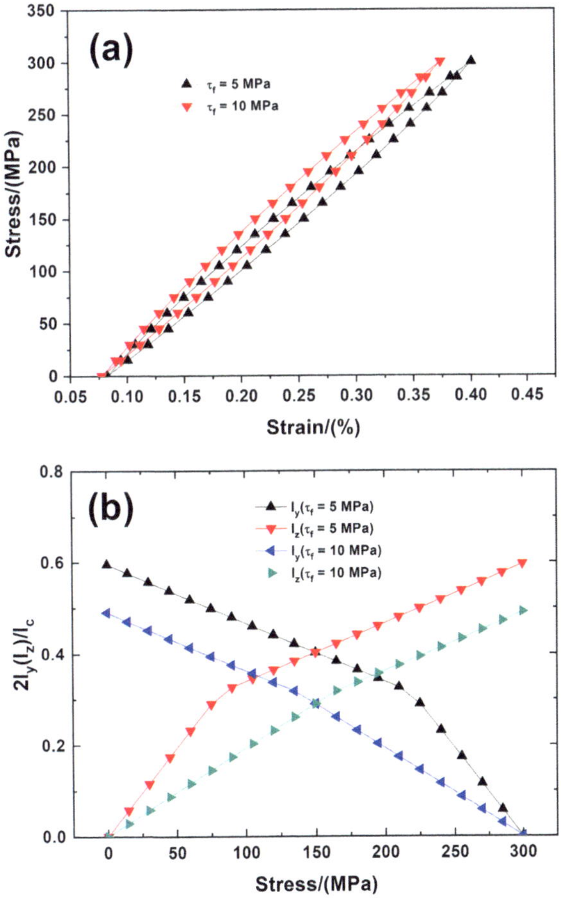

Fig. 2.9 Effect of interface shear stress in the wear region (i.e., $\tau_f = 5$ and 10 MPa) on **a** the fatigue hysteresis loops; **b** the interface slip ratio versus applied stress

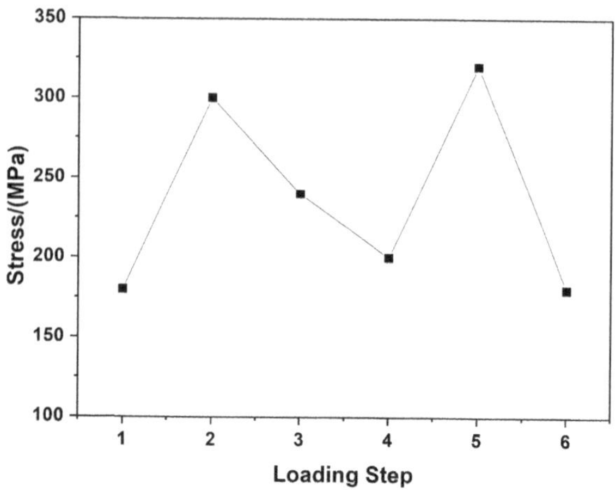

Fig. 2.10 Schematic of arbitrary loading stress levels

When the interface shear stress in the interface wear region increased, the unloading/reloading interface slip proportion, i.e., $2l_y/l_c$ and $2l_z/l_c$, decreased, leading to the decrease of unloading residual strain and fatigue hysteresis loops area, and the increase of fatigue hysteresis modulus.

2.3.6 Hysteresis Loops Under Stochastic Loading Sequence

Figure 2.10 shows the schematic of stochastic loading sequence, in which the loading sequence is given in Eq. (2.33), i.e., first loading at fatigue peak stress $\sigma_{\max 1} = 180$ MPa; and then considering interface wear in the debonding region, with the loading sequence $\sigma_{\max} = 300$, 240, 200, 320, and 280 MPa.

$$\begin{aligned} &\sigma_{\max 1}(180\text{MPa}) \rightarrow \sigma_{\max 2}(300\text{MPa}) \rightarrow \sigma_{\max 3}(240\text{MPa}) \\ &\quad \rightarrow \sigma_{\max 4}(200\text{MPa}) \rightarrow \sigma_{\max 5}(320\text{MPa}) \rightarrow \sigma_{\max 6}(180\text{MPa}) \end{aligned} \tag{2.33}$$

When increasing fatigue peak stress from $\sigma_{\max 1} = 180$ to $\sigma_{\max 2} = 300$ MPa, the fatigue hysteresis dissipated energy increases from $U = 40.6$ kJ/m^3 without considering interface wear to $U = 55.3$ kJ/m^3 considering interface wear, as shown in Fig. 2.11a; the unloading interface counter-slip length $2l_y/l_c$ and reloading interface new-slip length $2l_z/l_c$ increases from 38.5% without considering interface wear to 59.5% considering interface wear, as shown in Fig. 2.11b.

When decreasing fatigue peak stress from $\sigma_{\max 2} = 300$ to $\sigma_{\max 3} = 240$ MPa, the fatigue hysteresis loss energy increases from $U = 20.7$ kJ/m^3 without considering interface wear to $U = 31.9$ kJ/m^3 considering interface wear, as shown in Fig. 2.12a; the unloading interface counter-slip length $2l_y/l_c$ and reloading interface new-slip length $2l_z/l_c$ increases from 30.8% without considering interface wear to 51.8% considering interface wear, as shown in Fig. 2.12b.

Fig. 2.11 **a** Fatigue hysteresis loops; **b** the interface slip ratio versus applied stress with and without considering interface wear at $\sigma_{max2} = 300$ MPa

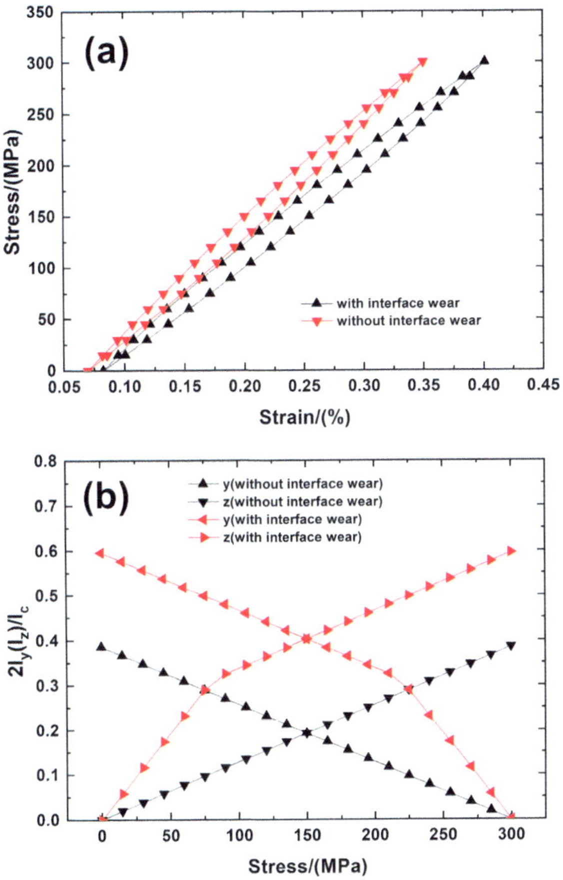

When decreasing fatigue peak stress from $\sigma_{max3} = 240$ to $\sigma_{max4} = 200$ MPa, the fatigue hysteresis loss energy increases from $U = 12$ kJ/m^3 without considering interface wear to $U = 20.7$ kJ/m^3 considering interface wear, as shown in Fig. 2.13a; the unloading interface counter-slip length $2l_y/l_c$ and reloading interface new-slip length $2l_z/l_c$ increases from 25.7% without considering interface wear to 46.7% considering interface wear, as shown in Fig. 2.13b.

When increasing fatigue peak stress from $\sigma_{max4} = 200$ to $\sigma_{max5} = 320$ MPa, the fatigue hysteresis loss energy increases from $U = 49.3$ kJ/m^3 without considering interface wear to $U = 65.2$ kJ/m^3 considering interface wear, as shown in Fig. 2.14a; the unloading interface counter-slip length $2l_y/l_c$ and reloading interface new-slip length $2l_z/l_c$ increases from 41.1% without considering interface wear to 62.1% considering interface wear, as shown in Fig. 2.14b.

When decreasing fatigue peak stress from $\sigma_{max5} = 320$ to $\sigma_{max6} = 180$ MPa, the fatigue hysteresis dissipated energy increases from $U = 8.7$ kJ/m^3 without considering interface wear to $U = 16.3$ kJ/m^3 considering interface wear, as shown in

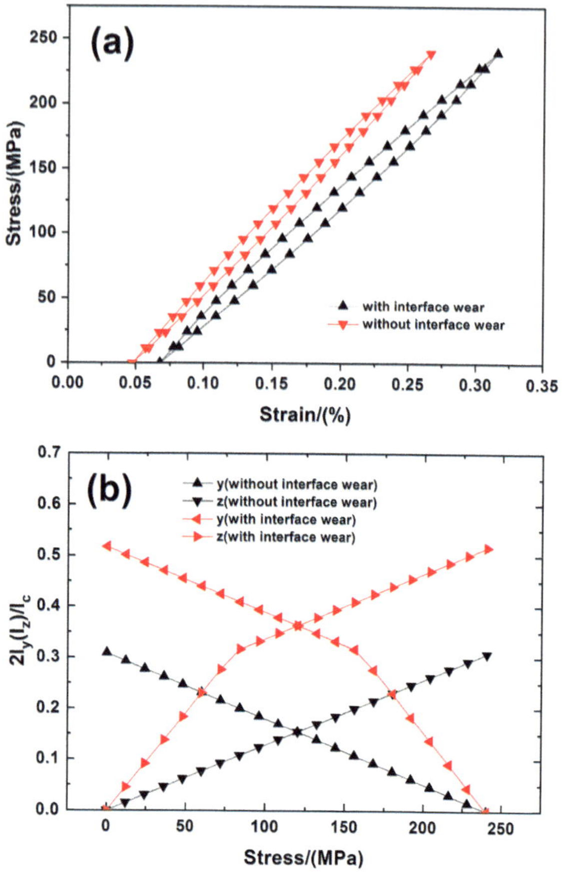

Fig. 2.12 **a** Fatigue hysteresis loops; **b** the interface slip ratio versus applied stress with and without considering interface wear at $\sigma_{max3} = 240$ MPa

Fig. 2.15a; the unloading interface counter-slip length $2l_y/l_c$ and reloading interface new-slip length $2l_z/l_c$ increases from 23.1% without considering interface wear to 44% considering interface wear, as shown in Fig. 2.15b.

2.4 Conclusion

In this chapter, the synergistic effect of stochastic loading sequence and interface wear on the fatigue hysteresis loops of fiber-reinforced CMCs was investigated. Under fatigue loading, fiber/matrix interface shear stress decreased with the increase of cycle number due to interface wear. Upon increasing fatigue peak stress, the interface debonding length propagated along the fiber/matrix interface. The difference of interface shear stress existing in the new and original interface debonding region affected the interface debonding and interface frictional sliding between the fiber

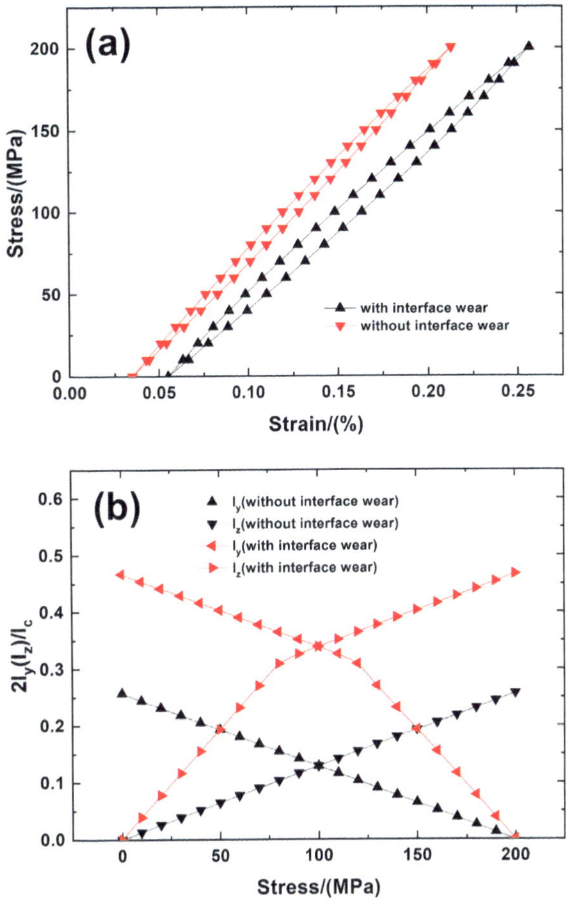

Fig. 2.13 **a** Fatigue hysteresis loops; **b** the interface slip ratio versus applied stress with and without considering interface wear at $\sigma_{\max 4} = 200$ MPa

and the matrix. Based on the fatigue damage mechanism of fiber sliding relative to matrix in the interface debonding region upon unloading and subsequent reloading, the interface debonding length, unloading interface counter-slip length, and reloading interface new-slip length were determined using the fracture mechanics approach. The fatigue hysteresis loops models considering the effect of stochastic fatigue stress levels and interface wear were developed. Effects of fatigue peak stress, fiber volume, matrix crack spacing, interface debonding, interface wear, and stochastic loading sequence on unloading/reloading interface slip and fatigue hysteresis loops were analyzed.

(1) When the fatigue peak stress increased, the unloading/reloading interface slip proportion increased, leading to the increase of unloading residual strain and fatigue hysteresis loops area, and the decrease of fatigue hysteresis modulus.

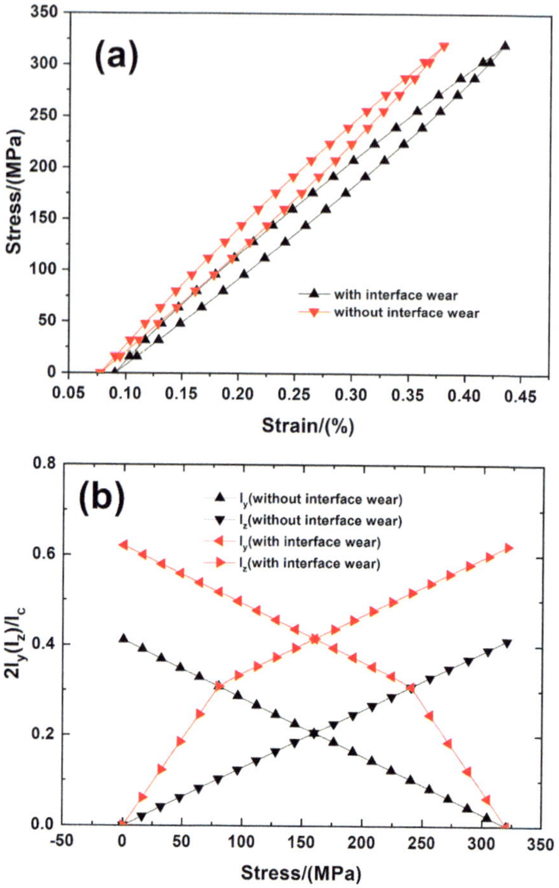

Fig. 2.14 **a** Fatigue hysteresis loops; **b** the interface slip ratio versus applied stress with and without considering interface wear at $\sigma_{\max 2} = 320$ MPa

When the fiber volume, matrix crack spacing, interface shear stress, and interface debonding energy increased, the unloading/reloading interface slip proportion decreased, leading to the decrease of unloading residual strain and fatigue hysteresis loops area, and the increase of fatigue hysteresis modulus.

(2) When the interface shear stress in the interface wear region increased, the unloading/reloading interface slip proportion decreased, leading to the decrease of unloading residual strain and fatigue hysteresis loops area, and the increase of fatigue hysteresis modulus.

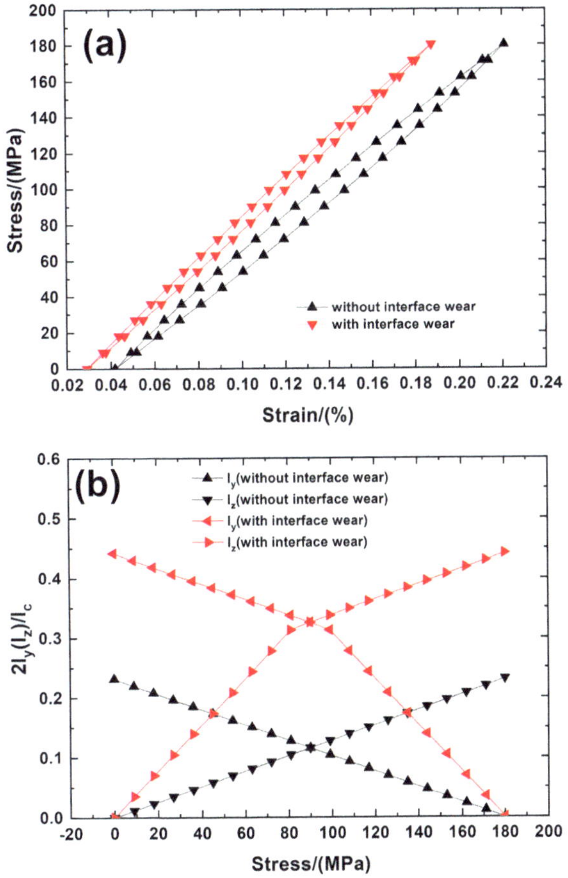

Fig. 2.15 **a** Fatigue hysteresis loops; **b** the interface slip ratio versus applied stress with and without considering interface wear at $\sigma_{\max 6} = 180$ MPa

References

Ahn BK, Curtin WA (1997) Strain and hysteresis by stochastic matrix cracking in ceramic matrix composites. J Mech Phys Solids 45:177–209. https://doi.org/10.1016/S0022-5096(96)00081-6

Budiansky B, Hutchinson JW, Evans AG (1986) Matrix fracture in fiber-reinforced ceramics. J Mech Phys Solids 34:167–189. https://doi.org/10.1016/0022-5096(86)90035-9

Evans AG, Zok FW, McMeeking RM (1995) Fatigue of ceramic matrix composites. Acta Metall Mater 43:859–875. https://doi.org/10.1016/0956-7151(94)00304-Z

Fantozzi G, Reynaud P (2009) Mechanical hysteresis in ceramic matrix composites. Mater Sci Eng A 521–522:18–23. https://doi.org/10.1016/j.msea.2008.09.128

Gao YC, Mai Y, Cotterell B (1988) Fracture of fiber-reinforced materials. J Appl Math Phys 39:550–572. https://doi.org/10.1007/BF00948962

Holmes JW, Cho CD (1992) Experimental observation of frictional heating in fiber-reinforced ceramics. J Am Ceram Soc 75:929–938. https://doi.org/10.1111/j.1151-2916.1992.tb04162.x

Kotil T, Holmes JW, Comninou M (1990) Origin of hysteresis observed during fatigue of ceramic-matrix composites. J Am Ceram Soc 73:1879–1883. https://doi.org/10.1111/j.1151-2916.1990.tb05239.x

Li LB (2013a) Modeling hysteresis behavior of cross-ply C/SiC ceramic matrix composites. Compos Part B Eng 53:36–45. https://doi.org/10.1016/j.compositesb.2013.04.029
Li LB (2013b) Fatigue hysteresis behavior of cross-ply C/SiC ceramic matrix composites at room and elevated temperatures. Mater Sci Eng A 586:160–170. https://doi.org/10.1016/j.msea.2013.08.017
Li LB (2014) Modeling fatigue hysteresis behavior of unidirectional C/SiC ceramic-matrix composite. Compos Part b: Eng 66:466–474. https://doi.org/10.1016/j.compositesb.2014.06.014
Li LB (2015b) Fatigue hysteresis of carbon fiber-reinforced ceramic-matrix composites at room and elevated temperatures. Appl Compos Mater 23:1–27. https://doi.org/10.1007/s10443-015-9448-1
Li LB (2015c) Modeling for fatigue hysteresis loops of carbon fiber-reinforced ceramic-matrix composites under multiple loading stress levels. Appl Compos Mater 22:945–959. https://doi.org/10.1007/s10443-015-9444-5
Li LB (2015d) Synergistic effect of arbitrary loading sequence and interface wear on the fatigue hysteresis loops of carbon fiber-reinforced ceramic-matrix composites. Eng Fract Mech 146:67–88. https://doi.org/10.1016/j.engfracmech.2015.07.060
Li LB (2015a) Micromechanics modelling of fatigue hysteresis loops in carbon fibre-reinforced ceramic-matrix composites. J Compos Mater 49:3471–3495. https://doi.org/10.1177/0021998314566055
Li LB (2019a) Thermomechanical fatigue of ceramic-matrix composites. Wiley-VCH, Weinheim, Germany
Li LB (2019b) Modeling matrix multi-cracking evolution of fiber-reinforced ceramic-matrix composites considering fiber fracture. Ceram-Silik 63:21–31
Li LB (2019c) Temperature-dependent proportional limit stress of carbon fiber-reinforced silicon carbide ceramic-matrix composites. Ceram-Silik 63:330–337
Li LB (2018) Damage, fracture and fatigue of ceramic-matrix composites. Springer Nature, Singapore. https://www.springer.com/in/book/9789811317828
Li LB (2020) Durability of ceramic-matrix composites. Woodhead Publishing, Oxford, UK. https://doi.org/10.1016/C2018-0-03346-0
Li LB, Song YD (2011) Effect of fiber failure on quasi-static unloading/reloading hysteresis loops of ceramic matrix composites. Trans Nanjing Uni Aero Astronaut 28:94–102
Li LB, Song YD, Sun ZG (2009a) Influence of interface de-bonding on the fatigue hysteresis loops of ceramic matrix composites. Chin J Solids Mech 30:8–14
Li LB, Song YD, Sun ZG, Sun J (2009b) Effect of fiber Poisson contraction on fatigue hysteresis loops of ceramic matrix composites. J Nanjing Uni Aero Astronaut 41:181–186
Naslain R (2004) Design, preparation and properties of non-oxide CMCs for application in engines and nuclear reactors: an overview. Compos Sci Technol 64:155–170. https://doi.org/10.1016/S0266-3538(03)00230-6
Pryce AW, Smith PA (1993) Matrix cracking in unidirectional ceramic matrix composites under quasi-static and cyclic loading. Acta Metall Mater 41:1269–1281. https://doi.org/10.1016/0956-7151(93)90178-U
Reynaud P (1996) Cyclic fatigue of ceramic-matrix composites at ambient and elevated temperatures. Compos Sci Technol 56:809–814. https://doi.org/10.1016/0266-3538(96)00025-5
Solti JP, Mall S, Robertson DD (1995) Modeling damage in unidirectional ceramic-matrix composites. Compos Sci Technol 54:55–66. https://doi.org/10.1016/0266-3538(95)00041-0
Solti JP, Mall S, Robertson DD (1997) Modeling of fatigue in cross-ply ceramic matrix composites. J Compos Mater 31:1921–1943. https://doi.org/10.1177/002199839703101903
Sun YJ, Singh RN (1998) The generation of multiple matrix cracking and fiber-matrix interfacial debonding in a glass composite. Acta Mater 46:1657–1667. https://doi.org/10.1016/S1359-6454(97)00347-9
Vagaggini E, Domergue JM, Evans AG (1995) Relationships between hysteresis measurements and the constituent properties of ceramic matrix composites: I. Theor J Am Ceram Soc 78:2709–2720. https://doi.org/10.1111/j.1151-2916.1995.tb08046.x

Wang Y, Zhang L, Cheng L, Mei H, Ma J (2008) Characterization of tensile behavior of a two-dimensional woven carbon/silicon carbide composite fabricated by chemical vapor infiltration. Mater Sci Eng A 497:295–300. https://doi.org/10.1016/j.msea.2008.07.050

Chapter 3
Stress-Rupture of Ceramic-Matrix Composites Under Stochastic Loading at Intermediate Temperature

Abstract In this chapter, stress-rupture of fiber-reinforced ceramic-matrix composites (CMCs) with stochastic loading at intermediate temperature is investigated. Four different stochastic loading sequences under stress-rupture are considered with a different loading time and time interval. Microstress field of the damaged CMCs and different damage models consider the effect of stochastic loading and time-dependent interface and fiber oxidation. Damage evolution of stress-rupture strain, interface debonding and interface oxidation ratio, and the broken fibers fraction versus time curves of SiC/SiC composite subjected to four different stochastic loading sequences are analyzed. Effects of composite constituent properties, damage state, and environmental temperature on damage evolution and lifetime of SiC/SiC composite are discussed. Experimental damage evolution and lifetime of SiC/SiC composite under stress-rupture with stochastic loading are predicted.

Keywords Ceramic-matrix composites (CMCs) · Stress-rupture · Stochastic loading · Lifetime

3.1 Introduction

Applications of fiber-reinforced CMCs on intermediate temperature (i.e., 700–1000 °C) and intermediate load (i.e., less than 120 MPa) stationary parts involve lives measured in tens of thousands of hours. It is necessary to investigate the stress-rupture behavior of fiber-reinforced CMCs at intermediate temperatures (Recle et al. 2017; Li 2020). Damage mechanisms of matrix cracking, interface debonding and oxidation, and fibers fracture affect the damage evolution and lifetime of CMCs under stress-rupture loading (Ruggles-Wrenn et al. 2011; Dassios et al. 2013; Rebillat 2014). Lara-Curzio (1997) investigated the stress-rupture of SiC/SiC composite at $T = 950$ °C in air atmosphere. The oxidation-induced stress-rupture of the reinforcing fiber bundles caused the delayed failure of SiC/SiC composite at elevated temperature. Morscher (1997) investigated the tensile stress-rupture behavior of SiC/SiC minicomposites with carbon and BN interphases at elevated temperature range of $T = 700$–1200 °C in air atmosphere. It was found that severe embrittlement occurred for the SiC/SiC minicomposite with the carbon interphase due to the degradation of

L. Li, *Damage and Fracture of Ceramic-Matrix Composites Under Stochastic Loading*, Advanced Ceramics and Composites 2,
https://doi.org/10.1007/978-981-16-2141-3_3

the Nicalon™ fiber. Verrilli et al. (2004) investigated the effect of environment on the stress-rupture behavior of C/SiC composite in air, vacuum, and steam-containing environment at elevated temperature of $T = 600$ and 1200 °C. The oxidation rate of the fibers in the various environments correlated with the stress-rupture lifetime of the C/SiC composite. Guathier and Lamon (2009) investigated the delayed failure of Hi-Nicalon and Hi-Nicalon S multifilament tows and single filaments at temperature range of $T = 500$–800 °C. A model based on slow crack growth in single filament shows that the stress-rupture of fiber tow follows the conventional time power law. Halverson and Curtin (2002) developed a micromechanically based model for the deformation, strength, and stress-rupture lifetime of a Nextel™ 610/alumina–yttria composite. The model considered the statistics of fiber strength, fiber degradation with time at temperature and load, the matrix damage, and the effects of fiber pullout within a global load sharing criterion. Maillet et al. (2012) and Godin et al. (2019) developed approaches based on the acoustic emission to predict the damage evolution and lifetime of SiC/SiC composite at intermediate temperature. Li (2017) investigated the damage evolution of cross-ply SiC/MAS composite under stress-rupture and cyclic loading at elevated temperature. The evolution of the strain and fiber/matrix interface debonding and oxidation versus the time is analyzed. Ikarashi et al. (2019) investigated the effect of cyclic loading on the stress-rupture behavior of 3D SiC/SiC composite at elevated temperature in air atmosphere. The oxidative matrix crack propagation and the decrease of the interfacial shear stress affect the lifetime of the SiC/SiC composite.

In this chapter, damage evolution and lifetime of fiber-reinforced CMCs under stress-rupture with stochastic loading at intermediate temperature are investigated. Damage evolution of composite strain, interface debonding and oxidation ratio, and the broken fibers fraction versus time curves of SiC/SiC composite subjected to constant stress and three different stochastic loading spectrums are analyzed. Effects of composite constituent properties, damage state, and environmental temperature on damage evolution and lifetime are discussed. Experimental damage evolution and lifetime of SiC/SiC composite under stress-rupture with stochastic loading are predicted.

3.2 Theoretical Model

When stochastic loading occurs under constant stress loading at elevated temperature, the damage extent inside of fiber-reinforced CMCs becomes much more serious. In the present analysis, the shear-lag model is used to analyze the stress distribution of damaged CMCs under stress-rupture with stochastic loading. Damage mechanisms of matrix cracking, interface debonding and oxidation, and fibers broken are considered. The matrix stochastic cracking model, fracture mechanics approach, and global load sharing (GLS) criterion are used to determine the matrix crack spacing, interface debonding ratio, and the broken fibers fraction under stress-rupture with stochastic loading. Figure 3.1 shows the stochastic loading spectrum under constant stress-rupture loading of fiber-reinforced CMCs at elevated temperature, which can be divided into four cases, as follows:

(1) Case 1, constant stress loading.
(2) Case 2, constant stress loading and stochastic loading σ_a with Δt_a.

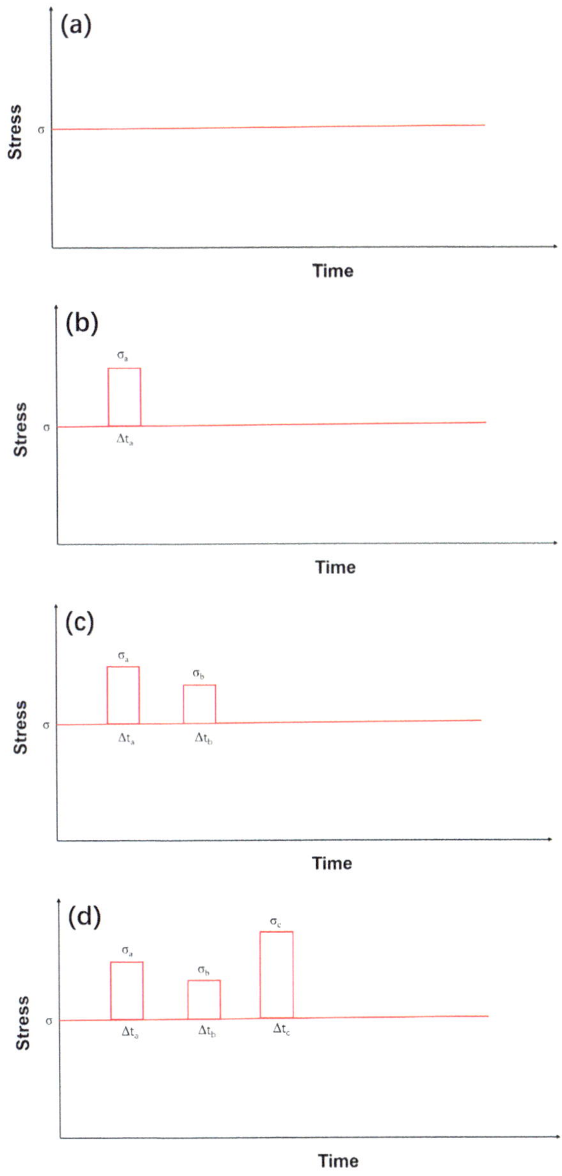

Fig. 3.1 Schematic of loading spectrum of **a** constant loading; **b** stochastic loading σ_a and Δt_a; **c** stochastic loading σ_a, σ_b and Δt_a, Δt_b; and **d** stochastic loading σ_a, σ_b, σ_c and Δt_a, Δt_b, Δt_c

(3) Case 3, constant stress loading and stochastic loading σ_a and σ_b with Δt_a and Δt_b.
(4) Case 4, constant stress loading and stochastic loading σ_a, σ_b, and σ_c with Δt_a, Δt_b, and Δt_c.

When fibers fracture occurs under stochastic loading, the fiber axial stress distributions in the interface oxidation region, interface slip region, and the interface bonding region are

$$\sigma_f(x,t) = \begin{cases} \Phi_s(t) - \frac{2\tau_f}{r_f}x,\ x \in [0, \zeta(t)] \\ \Phi_s(t) - \frac{2\tau_f}{r_f}\zeta(t) - \frac{2\tau_i}{r_f}(x - \zeta(t)),\ x \in [\zeta(t), l_d(t)] \\ \sigma_{fo} + \left[\Phi_s(t) - \sigma_{fo} - \frac{2\tau_f}{r_f}\zeta(t) - \frac{2\tau_i}{r_f}(l_d(t) - \zeta(t))\right] \\ \exp\left(-\rho\frac{x - l_d(t)}{r_f}\right),\ x \in \left[l_d(t), \frac{l_c}{2}\right] \end{cases} \tag{3.1}$$

where r_f is the fiber radius; τ_f is the interface shear stress in the oxidation region; τ_i is the interface shear stress in the slip region; $\Phi_s(t)$ is the intact fiber stress under stochastic loading; $l_d(t)$ is the time-dependent interface debonding length under stochastic loading; l_c is the matrix crack spacing under stochastic loading; ρ is the shear-lag model parameter; and $\zeta(t)$ is the time-dependent fiber/matrix interface oxidation length (Casas and Martinez-Esnaola 2003).

$$\zeta(t) = \varphi_1\left[1 - \exp\left(-\frac{\varphi_2 t}{b}\right)\right] \tag{3.2}$$

where b is a delay factor considering the deceleration of reduced oxygen activity, and φ_1 and φ_2 are parameters dependent on temperature and described using the Arrhenius-type laws. The fiber axial stress in the fiber/matrix interface bonded region is

$$\sigma_{fo} = \frac{E_f}{E_c}\sigma + E_f(\alpha_c - \alpha_f)\Delta T \tag{3.3}$$

where E_f and E_c denote the fiber and the composite elastic modulus, respectively; α_f and α_c denote the fiber and the composite thermal expansion coefficient, respectively; and ΔT denotes the temperature difference between the testing temperature and the fabrication temperature.

Under stress-rupture with stochastic loading, stochastic stress affects matrix crack spacing. The matrix cracking under stochastic loading can be described using the two-parameter Weibull distribution, and the time/stress-dependent matrix crack spacing under stochastic loading can be determined using the following equation.

$$l_c(\sigma_s, t) = r_f\frac{V_m E_m}{V_f E_c}\frac{\sigma_R}{2\tau_i}\Delta\left\{1 - \exp\left[-\left(\frac{\sigma_s - (\sigma_{mc} - \sigma_{th})}{(\sigma_R - \sigma_{th}) - (\sigma_{mc} - \sigma_{th})}\right)^m\right]\right\}^{-1} \tag{3.4}$$

where σ_s is the stochastic loading stress, E_m denotes the matrix elastic modulus, σ_R denotes the matrix cracking characteristic strength, σ_{mc} denotes matrix first cracking stress, σ_{th} denotes matrix thermal residual stress, Δ denotes the final nominal crack space, and m denotes matrix Weibull modulus.

Under stress-rupture with stochastic loading, the interface debonding length is affected by stochastic loading. Time-dependent fiber/matrix interface debonding length under stochastic loading can be determined using the fracture mechanics approach (Gao et al. 1988).

$$\Gamma_d = -\frac{F}{4\pi r_f}\frac{\partial w_f(\sigma_s, t)}{\partial l_d} - \frac{1}{2}\int_0^{l_d} \tau_i \frac{\partial v(\sigma_s, t)}{\partial l_d} dx \tag{3.5}$$

where Γ_d denotes the interface debonding energy, $F(=\pi r_f^2 \sigma / V_f)$ is the fiber stress on the matrix cracking plane, $w_f(\sigma_s, t)$ denotes the time-dependent fiber axial displacement under stochastic loading at the matrix cracking plane, and $v(\sigma_s, t)$ denotes the time-dependent relative displacement between the fiber and the matrix under stochastic loading. Substituting the time-dependent fiber axial displacement and relative displacement into Eq. (3.5), the time-dependent fiber/matrix interface debonding length under stochastic loading can be determined using the following equation.

$$l_d(\sigma_s, t) = \left(1 - \frac{\tau_f}{\tau_i}\right)\zeta(t) + \frac{r_f}{2}\left(\frac{V_m E_m \Phi_s}{E_c \tau_i} - \frac{1}{\rho}\right) - \sqrt{\left(\frac{r_f}{2\rho}\right)^2 - \frac{r_f^2 V_f V_m E_f E_m \Phi_s^2}{4E_c^2 \tau_i^2}\left(1 - \frac{\sigma}{V_f \Phi_s}\right) + \frac{r_f V_m E_f E_m}{E_c \tau_i^2}\Gamma_d} \tag{3.6}$$

Under stress-rupture with stochastic loading, fiber fracture is affected by stochastic loading. Two-parameter Weibull model is adopted to describe fiber strength distribution, and the global load sharing (GLS) criterion is used to determine the stress distributions between the intact and fracture fibers (Curtin 1991).

$$\frac{\sigma_s}{V_f} = \Phi_s(1 - q(\Phi_s)) + \frac{2\tau_f}{r_f}\langle L \rangle q(\Phi_s) \tag{3.7}$$

where $\langle L \rangle$ denotes the average fiber pullout length and $q(\Phi_s)$ is the fiber failure probability.

$$q(\Phi_s) = 1 - \exp\left[-\left(\frac{\Phi_s}{\sigma_{fc}}\right)^{m_f+1}\right] \tag{3.8}$$

where m_f denotes the fiber Weibull modulus and σ_{fc} denotes the fiber characteristic strength of a length δ_c of fiber.

$$\sigma_{\mathrm{fc}} = \left(\frac{l_{\mathrm{o}}\sigma_0^{m_{\mathrm{f}}}(t)\tau_{\mathrm{i}}}{r_{\mathrm{f}}}\right)^{\frac{1}{m_{\mathrm{f}}+1}}, \delta_{\mathrm{c}} = \left(\frac{\sigma_0(t) r_{\mathrm{f}} l_{\mathrm{o}}^{\frac{1}{m_{\mathrm{f}}}}}{\tau_{\mathrm{i}}}\right)^{\frac{m_{\mathrm{f}}}{m_{\mathrm{f}}+1}} \tag{3.9}$$

where (Lara-Curzio 1999)

$$\sigma_0(t) = \begin{cases} \sigma_0, & t \le \frac{1}{k}\left(\frac{K_{\mathrm{IC}}}{Y\sigma_0}\right)^4 \\ \frac{K_{\mathrm{IC}}}{Y\sqrt[4]{kt}}, & t > \frac{1}{k}\left(\frac{K_{\mathrm{IC}}}{Y\sigma_0}\right)^4 \end{cases} \tag{3.10}$$

where σ_0 denotes the fiber strength; K_{IC} denotes the fracture toughness; Y denotes the geometric parameter; and k is the parabolic rate constant.

When multiple damage mechanisms form inside of fiber-reinforced CMCs under stress-rupture with stochastic loading, the average strain of the composite of $\varepsilon_{\mathrm{c}}(t)$ can be determined by integration of the axial strain in the fiber.

$$\varepsilon_{\mathrm{c}}(\sigma_{\mathrm{s}}, t) = \frac{2}{E_{\mathrm{f}} l_{\mathrm{c}}} \int\limits_{l_{\mathrm{c}}/2} \sigma_{\mathrm{f}}(x, t) dx - (\alpha_{\mathrm{c}} - \alpha_{\mathrm{f}})\Delta T \tag{3.11}$$

Substituting time-dependent fiber axial stress under stochastic loading in Eq. (3.1) into Eq. (3.11), the composite average strain of $\varepsilon_{\mathrm{c}}(\sigma_{\mathrm{s}}, t)$ can be determined using the following equation.

$$\varepsilon_{\mathrm{c}}(\sigma_{\mathrm{s}}, t) = \begin{cases} \frac{\Phi_{\mathrm{s}}(t)}{E_{\mathrm{f}}}\frac{2l_{\mathrm{d}}(t)}{l_{\mathrm{c}}} + \frac{2\tau_{\mathrm{f}}}{r_{\mathrm{f}}E_{\mathrm{f}}l_{\mathrm{c}}}\zeta^2(t) - \frac{4\tau_{\mathrm{f}}l_{\mathrm{d}}(t)}{r_{\mathrm{f}}E_{\mathrm{f}}l_{\mathrm{c}}}\zeta(t) - \frac{2\tau_{\mathrm{i}}}{r_{\mathrm{f}}E_{\mathrm{f}}l_{\mathrm{c}}}(l_{\mathrm{d}}(t) - \zeta(t))^2 \\ + \frac{2\sigma_{\mathrm{fo}}}{E_{\mathrm{f}}l_{\mathrm{c}}}\left(\frac{l_{\mathrm{c}}}{2} - l_{\mathrm{d}}(t)\right) + \frac{2r_{\mathrm{f}}}{\rho E_{\mathrm{f}}l_{\mathrm{c}}}\left\{\Phi_s(t) - \frac{2\tau_{\mathrm{f}}}{r_{\mathrm{f}}}\zeta(t) - \frac{2\tau_{\mathrm{i}}}{r_{\mathrm{f}}}[l_{\mathrm{d}}(t) - \zeta(t)] - \sigma_{\mathrm{fo}}\right\} \\ \times\left[1 - \exp\left(-\rho\frac{l_{\mathrm{c}}/2 - l_{\mathrm{d}}(t)}{r_{\mathrm{f}}}\right)\right] - (\alpha_{\mathrm{c}} - \alpha_{\mathrm{f}})\Delta T, l_{\mathrm{d}}(t) < \frac{l_{\mathrm{c}}}{2} \\ \frac{\Phi_{\mathrm{s}}(t)}{E_{\mathrm{f}}}\frac{2l_{\mathrm{d}}(t)}{l_{\mathrm{c}}} + \frac{2\tau_{\mathrm{f}}}{r_{\mathrm{f}}E_{\mathrm{f}}l_{\mathrm{c}}}\zeta^2(t) - \frac{4\tau_{\mathrm{f}}l_{\mathrm{d}}(t)}{r_{\mathrm{f}}E_{\mathrm{f}}l_{\mathrm{c}}}\zeta(t) - \frac{2\tau_{\mathrm{i}}}{r_{\mathrm{f}}E_{\mathrm{f}}l_{\mathrm{c}}}(l_{\mathrm{d}}(t) - \zeta(t))^2, l_d(t) = \frac{l_{\mathrm{c}}}{2} \end{cases} \tag{3.12}$$

Under stress-rupture with stochastic loading, the fiber axial stress distribution in different damaged regions, the matrix crack spacing, the fiber/matrix interface debonding length, and the fiber failure probability are affected by the stochastic stress level. The matrix crack spacing is determined using stochastic matrix cracking model by Eq. (3.4); the fiber/matrix interface debonding length is determined using the fracture mechanics approach by Eq. (3.6); the intact fiber stress is determined using the global load sharing criterion by Eq. (3.7). The stress-rupture strain can be determined by Eq. (3.12), when the fiber failure probability approaches critical value, the composite fracture.

3.3 Result and Discussion

Figure 3.1 shows the stochastic loading spectrum under constant stress-rupture loading of fiber-reinforced CMCs at elevated temperature, which can be divided into four cases, as follows:

(1) Case 1, constant stress $\sigma = 120$ MPa.
(2) Case 2, constant stress $\sigma = 120$ MPa and stochastic stress $\sigma_s = 140$ MPa at $t = 36$ ks with $\Delta t = 36$ ks.
(3) Case 3, constant stress $\sigma = 120$ MPa and stochastic stress $\sigma_s = 140$ and 160 MPa at $t = 36$ and 108 kseconds with $\Delta t = 36$ ks.
(4) Case 4, constant stress $\sigma = 120$ MPa and stochastic stress $\sigma_s = 140$, 160, and 180 MPa at $t = 36$, 108, and 180 ks with $\Delta t = 36$ ks.

Effects of composite constituent properties, damage state and environmental temperature on composite strain, interface debonding ratio, and interface oxidation ratio, broken fibers fraction versus time curves of SiC/SiC composite are analyzed for Cases 1, 2, 3, and 4. Material properties of SiC/SiC composite are given by: $V_f = 20\%$, $E_f = 270$ GPa, $E_m = 400$ GPa, $r_f = 7$ μm, $m = 3$, $\alpha_f = 3.5 \times 10^{-6}$/°C, $\alpha_m = 4.6 \times 10^{-6}$/°C, $\Delta T = -1000$ °C, $\Gamma_d = 0.1$ J/m^2, $\tau_i = 30$ MPa, $\tau_f = 1$ MPa, $\sigma_0 = 2.5$ GPa, $l_0 = 25$ mm, $m_f = 5$, and $T = 800$ °C.

3.3.1 *Stress-Rupture of SiC/SiC Composite with Stochastic Loading for Different Fiber Volumes*

Figures 3.2 and 3.3, and Tables 3.1 and 3.2 show the composite strain, interface debonding and oxidation ratio, and the broken fibers fraction of SiC/SiC composite corresponding to loading spectrums of Cases 1, 2, 3, and 4 at $T = 800$ °C in air atmosphere for $V_f = 0.2$ and 0.3. When the fiber volume increases, the stress-rupture lifetime increases, the time for the interface complete debonding increases, and the failure strain decreases. When fiber volume increases, the stress carried by intact fibers decreases, leading to the increase of the time for the complete interface debonding.

When the fiber volume is $V_f = 0.2$, the damage evolution and lifetime of SiC/SiC composite corresponding to loading spectrums of Cases 1, 2, 3, and 4 at $T = 800$ °C in air atmosphere are analyzed, as follows:

- Under constant stress $\sigma = 120$ MPa, the composite's stress-rupture lifetime is $t = 838.7$ ks; the time for the composite's interface complete debonding is $t = 213.6$ ks; the time for the composite's interface complete oxidation is $t = 295.3$ ks; the composite's failure strain is $\varepsilon_c = 0.253\%$; and the corresponding broken fibers fraction is $q = 0.285$.
- For the stochastic loading Case 2, the stochastic loading stress of $\sigma_s = 140$ MPa occurs at $t = 36$ ks; the time interval is $\Delta t = 36$ ks; the composite's stress-rupture

Fig. 3.2 **a** Composite strain versus the time curves; **b** the interface debonding ratio versus the time curves; **c** the interface oxidation ratio versus the time curves; and **d** the broken fibers fraction versus the time curves of SiC/SiC composite under stress-rupture loading of constant stress σ = 120 MPa, σ_s = 140, 140/160, 140/160/180 MPa at t = 36, 36/108, 36/108/180 ks and Δt = 36 ks at T = 800 °C in air atmosphere when V_f = 0.2

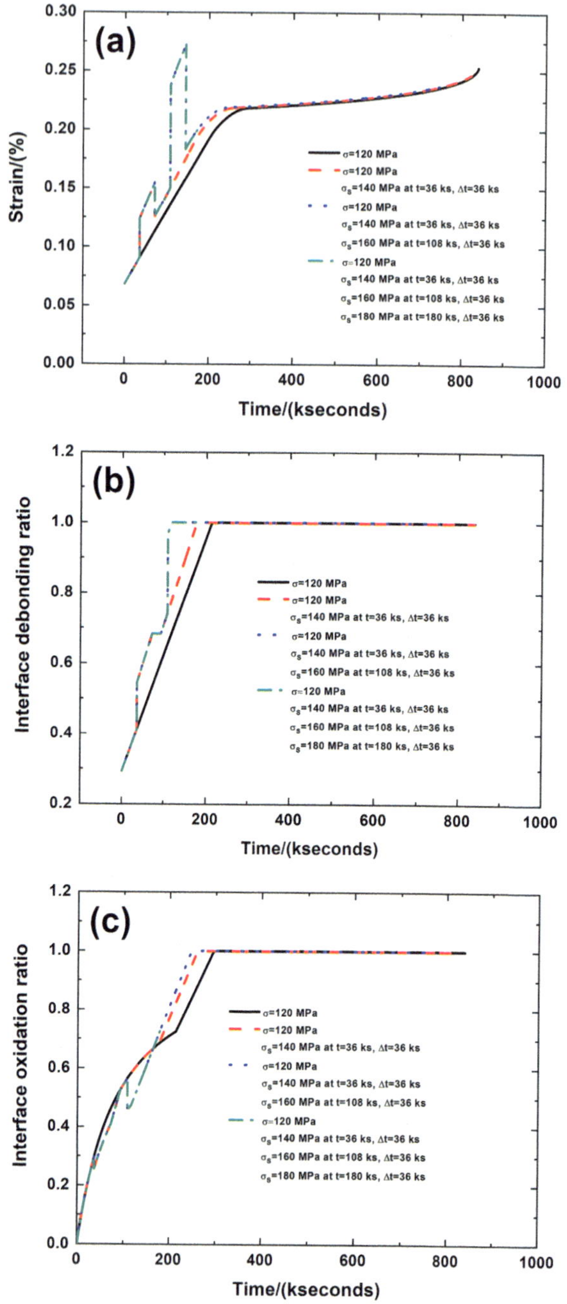

Fig. 3.2 (continued)

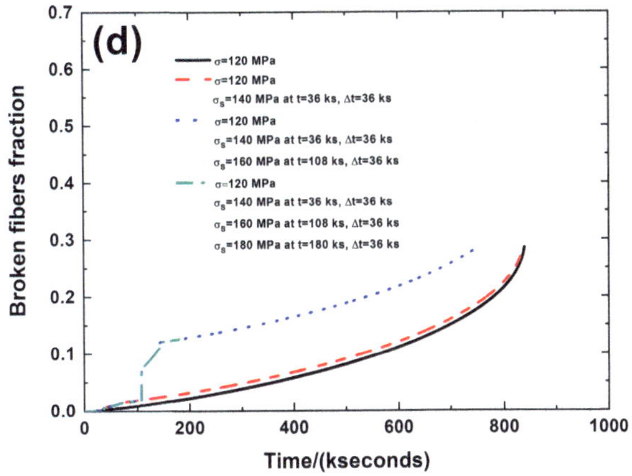

lifetime is $t = 837$ ks; the time for the composite's interface complete debonding is $t = 177$ ks under the constant stress $\sigma = 120$ MPa; the time for the composite's interface complete oxidation is t = 259.4 ks under constant stress $\sigma = 120$ MPa; the composite's failure strain is $\varepsilon_c = 0.253\%$; and the corresponding broken fibers fraction is $q = 0.285$.

- For the stochastic loading Case 3, the stochastic loading stresses of $\sigma_s = 140$, 160 MPa occur at $t = 36$, 108 ks; the time interval is $\Delta t = 36$ ks; the composite's stress-rupture lifetime is $t = 752.9$ ks; the time for the composite's interface complete debonding is $t = 116$ ks under stochastic loading stress of $\sigma_s = 160$ MPa; the time for the composite's interface complete oxidation is $t = 246.8$ ks under constant stress $\sigma = 120$ MPa; the composite's failure strain is $\varepsilon_c = 0.238\%$; and the corresponding broken fibers fraction is $q = 0.285$.
- For the stochastic loading Case 4, the stochastic loading stresses of $\sigma_s = 140$, 160, 180 MPa occur at $t = 36$, 108, 180 ks; the time interval is $\Delta t = 36$ ks; and the composite's stress-rupture lifetime is $t = 180$ ks upon increasing to the 3rd stochastic loading stress of $\sigma_s = 180$ MPa.

When the fiber volume is $V_f = 0.3$, the damage evolution and lifetime of SiC/SiC composite corresponding to loading spectrums of Cases 1, 2, 3, and 4 at $T = 800$ °C in air atmosphere are analyzed, as follows:

- Under constant stress $\sigma = 120$ MPa, the composite's stress-rupture lifetime is $t = 5873.2$ ks; the time for the composite's interface complete debonding is $t = 261.1$ ks; the time for the composite's interface complete oxidation is $t = 295.3$ ks; the composite's failure strain is $\varepsilon_c = 0.166\%$; and the corresponding broken fibers fraction is $q = 0.285$.
- For the stochastic loading Case 2, the stochastic loading stress of $\sigma_s = 140$ MPa occurs at $t = 36$ ks; the time interval is $\Delta t = 36$ ks; the composite's stress-rupture lifetime is $t = 5872.5$ ks; the time for the composite's interface complete debonding is $t = 224$ ks under the constant stress $\sigma = 120$ MPa; the time for the

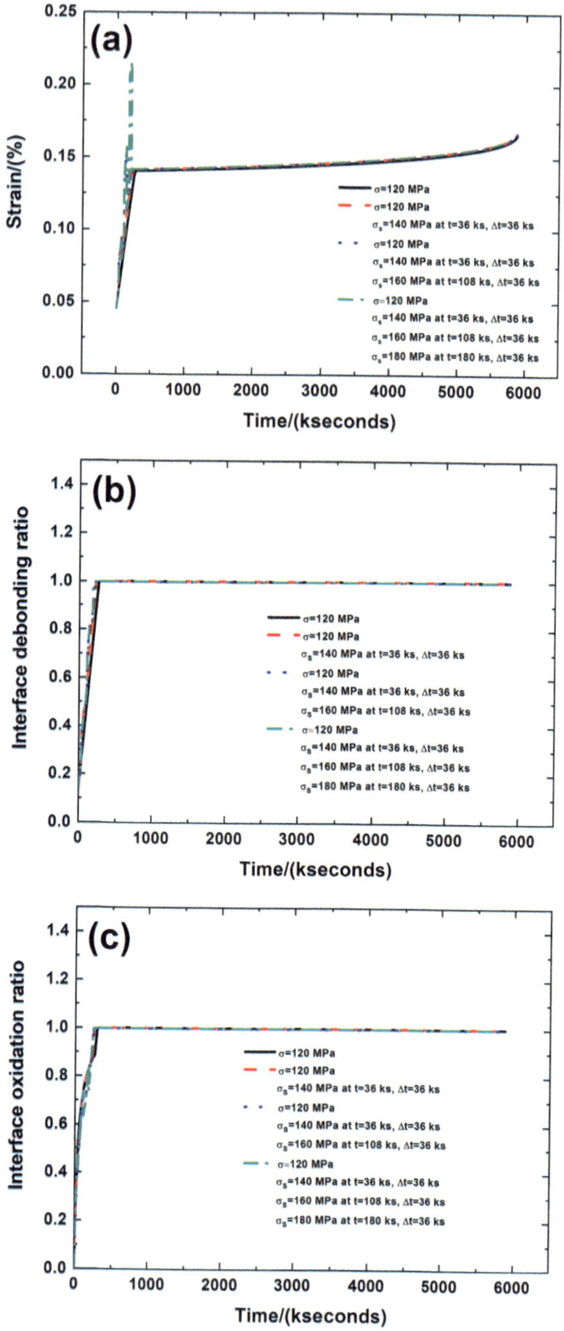

Fig. 3.3 **a** Composite strain versus the time curves; **b** the interface debonding ratio versus the time curves; **c** the interface oxidation ratio versus the time curves; and **d** the broken fibers fraction versus the time curves of SiC/SiC composite under stress-rupture loading of constant stress $\sigma =$ 120 MPa, $\sigma_s = 140$, 140/160, 140/160/180 MPa at $t = 36$, 36/108, 36/108/180 ks and $\Delta t =$ 36 ks at $T = 800$ °C in air atmosphere when $V_f = 0.3$

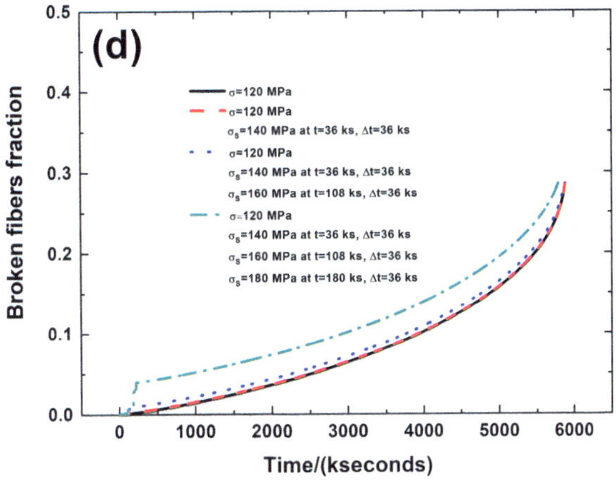

Fig. 3.3 (continued)

composite's interface complete oxidation is $t = 259.4$ ks under constant stress $\sigma = 120$ MPa; the composite's failure strain is $\varepsilon_c = 0.167\%$; and the corresponding broken fibers fraction is $q = 0.285$.

- For the stochastic loading Case 3, the stochastic loading stresses of $\sigma_s = 140$, 160 MPa occur at $t = 36$, 108 ks; the time interval is $\Delta t = 36$ ks; the composite's stress-rupture lifetime is $t = 5864.4$ ks; the time for the composite's interface complete debonding is $t = 211$ ks under constant stress $\sigma = 120$ MPa; the time for the composite's interface complete oxidation is $t = 246.8$ ks under constant stress $\sigma = 120$ MPa; the composite's failure strain is $\varepsilon_c = 0.167\%$; and the corresponding broken fibers fraction is $q = 0.285$.
- For the stochastic loading Case 4, the stochastic loading stresses of $\sigma_s = 140$, 160, 180 MPa occur at $t = 36$, 108, 180 ks; the time interval is $\Delta t = 36$ ks; the composite's stress-rupture lifetime is $t = 5783.7$ ks; the time for the composite's interface complete debonding is $t = 180$ ks under stochastic loading stress $\sigma_s = 180$ MPa; the time for the composite's interface complete oxidation is $t = 243.3$ ks under constant stress $\sigma = 120$ MPa; the composite's failure strain is $\varepsilon_c = 0.163\%$; and the corresponding broken fibers fraction is $q = 0.285$.

3.3.2 Stress-Rupture of SiC/SiC Composite with Stochastic Loading for a Different Matrix Crack Spacing

Figures 3.4 and 3.5, and Tables 3.3 and 3.4 show the composite's strain, interface debonding and oxidation ratio, and the broken fibers fraction of SiC/SiC composite corresponding to the loading sequences of Cases 1, 2, 3, and 4 at $T = 800$ °C in air atmosphere for $l_s = 100$ and 200 μm. When the saturation matrix crack spacing increases, the time for the interface complete debonding and oxidation increases and the failure strain decreases. The matrix crack spacing at the stress level below the

Table 3.1 Composite strain, interface debonding and oxidation ratio, and broken fibers fraction of SiC/SiC composite under stress-rupture loading of constant stress $\sigma = 120$ MPa, $\sigma_s = 140$, 140/160, 140/160/180 MPa at $t = 36$, 36/108, 36/108/180 ks and $\Delta t = 36$ ks at $T = 800$°C in air atmosphere when $V_f = 0.2$

σ/MPa	120	120	120	120	120					
t/ks	0	36	213.6	295.3	838.7					
ε_c/%	0.068	0.09	0.199	0.218	0.253					
η	0.293	0.412	1.0	1.0	1.0					
ω	0	0.297	0.724	1.0	1.0					
q	4×10^{-6}	0.002	0.024	0.037	0.285					
σ/MPa	120	120	140	140	120	120	120	120		
t/ks	0	36	36	72	72	177	259.4	837		
ε_c/%	0.068	0.09	0.124	0.154	0.126	0.196	0.218	0.253		
η	0.293	0.412	0.545	0.683	0.683	1.0	1.0	1.0		
ω	0	0.297	0.255	0.407	0.407	0.683	1.0	1.0		
q	4×10^{-6}	0.002	0.006	0.015	0.015	0.028	0.041	0.285		
σ/MPa	120	120	140	140	120	120	160	160	120	120	120
t/ks	0	36	36	72	72	108	108	144	144	246.8	752.9
ε_c/%	0.068	0.09	0.124	0.154	0.126	0.15	0.24	0.272	0.183	0.218	0.238
η	0.293	0.412	0.545	0.683	0.683	0.74	0.964	1.0	1.0	1.0	1.0
ω	0	0.297	0.255	0.407	0.407	0.564	0.454	0.584	0.584	1.0	1.0
q	4×10^{-6}	0.002	0.006	0.015	0.015	0.019	0.074	0.112	0.121	0.136	0.285
σ/MPa	120	120	140	140	120	120	160	160	120	120	180
t/ks	0	36	36	72	72	108	108	144	144	180	180
ε_c/%	0.068	0.09	0.124	0.154	0.126	0.15	0.24	0.272	0.183	0.203	–
η	0.293	0.412	0.545	0.683	0.683	0.74	0.964	1.0	1.0	1.0	–
ω	0	0.297	0.255	0.407	0.407	0.564	0.454	0.584	0.584	0.73	–
q	4×10^{-6}	0.002	0.006	0.015	0.015	0.019	0.074	0.112	0.121	0.126	–

saturation matrix crack stress increases with saturation matrix crack spacing, leading to the increase of the time for the interface complete debonding and oxidation.

When the saturation matrix crack spacing is $l_s = 100$ μm, the damage evolution and lifetime of SiC/SiC composite corresponding to loading spectrums of Cases 1, 2, 3, and 4 at $T = 800$ °C in air atmosphere are analyzed, as follows:

- Under constant stress $\sigma = 120$ MPa, the composite's stress-rupture lifetime is $t = 2447.9$ ks; the time for the composite's interface complete debonding is $t = 59.8$ ks; the time for the composite's interface complete oxidation is $t = 117.8$ ks; the composite's failure strain is $\varepsilon_c = 0.206\%$; and the corresponding broken fibers fraction is $q = 0.285$.

Table 3.2 Composite strain, interface debonding and oxidation ratio, and broken fibers fraction of SiC/SiC composite under stress-rupture loading of constant stress $\sigma = 120$ MPa, $\sigma_s = 140$, 140/160, 140/160/180 MPa at $t = 36$, 36/108, 36/108/180 kseconds and $\Delta t = 36$ ks at $T = 800$ °C in air atmosphere when $V_f = 0.3$

σ/MPa	120	120	120	120	120										
t/ks	0	36	261.1	295.3	5873.2										
ε_c/%	0.045	0.059	0.137	0.14	0.166										
η	0.144	0.262	1.0	1.0	1.0										
ω	0	0.466	0.884	1.0	1.0										
q	0	2.1×10^{-4}	0.0026	0.003	0.285										
σ/MPa	120	120	140	140	120	120	120	120							
t/ks	0	36	36	72	72	224	259.4	5872.5							
ε_c/%	0.045	0.059	0.076	0.094	0.079	0.137	0.141	0.167							
η	0.144	0.262	0.344	0.478	0.478	1.0	1.0	1.0							
ω	0	0.466	0.405	0.581	0.581	0.864	1.0	1.0							
q	0	2.1×10^{-4}	5.4×10^{-4}	0.001	0.001	0.003	0.0033	0.285							
σ/MPa	120	120	140	140	120	120	160	160	120	120	120	120			
t/ks	0	36	36	72	72	108	108	144	144	211	246.8	5864.4			
ε_c/%	0.045	0.059	0.076	0.094	0.079	0.092	0.136	0.157	0.112	0.137	0.141	0.167			
η	0.144	0.262	0.344	0.478	0.478	0.568	0.693	0.834	0.834	1.0	1.0	1.0			
ω	0	0.466	0.405	0.581	0.581	0.735	0.633	0.7	0.7	0.855	1.0	1.0			
q	0	2.1×10^{-4}	5.4×10^{-4}	0.001	0.001	0.0016	0.0056	0.0077	0.0085	0.009	0.0097	0.285			

(continued)

Table 3.2 (continued)

σ/MPa	120	120	140	140	120	120	160	160	120	120	180	180	120	120	120
t/ks	0	36	36	72	72	108	108	144	144	180	180	216	216	243.3	5783.7
ε_c/%	0.045	0.059	0.076	0.094	0.079	0.092	0.136	0.157	0.112	0.125	0.204	0.216	0.139	0.141	0.163
η	0.144	0.262	0.344	0.478	0.478	0.568	0.693	0.834	0.834	0.878	1.0	1.0	1.0	1.0	1.0
ω	0	0.466	0.405	0.581	0.581	0.735	0.633	0.7	0.7	0.831	0.74	0.888	0.888	1.0	1.0
q	0	2.1×10^{-4}	5.4×10^{-4}	0.001	0.001	0.0016	0.0056	0.0077	0.0085	0.0089	0.026	0.032	0.039	0.04	0.285

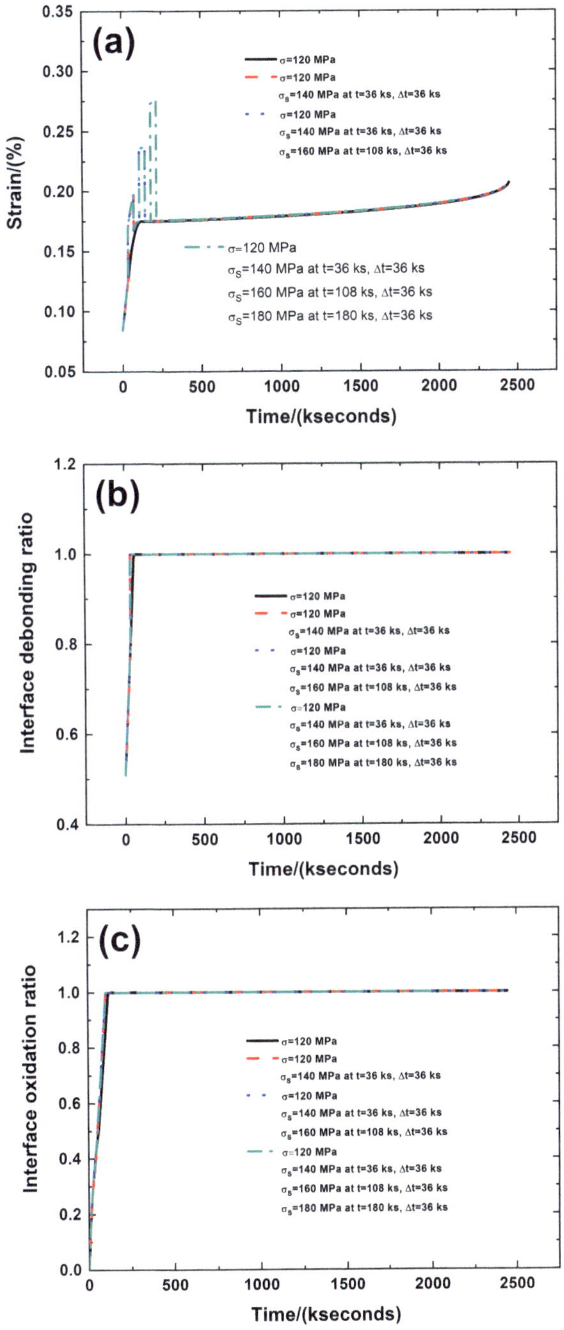

Fig. 3.4 **a** Composite strain versus the time curves; **b** the interface debonding ratio versus the time curves; **c** the interface oxidation ratio versus the time curves; and **d** the broken fibers fraction versus the time curves of SiC/SiC composite under stress-rupture loading of constant stress $\sigma = 120$ MPa, $\sigma_s = 140$, 140/160, 140/160/180 MPa at $t = 36$, 36/108, 36/108/180 ks and $\Delta t = 36$ ks at $T = 800$ °C in air atmosphere when $l_s = 100$ μm

Fig. 3.4 (continued)

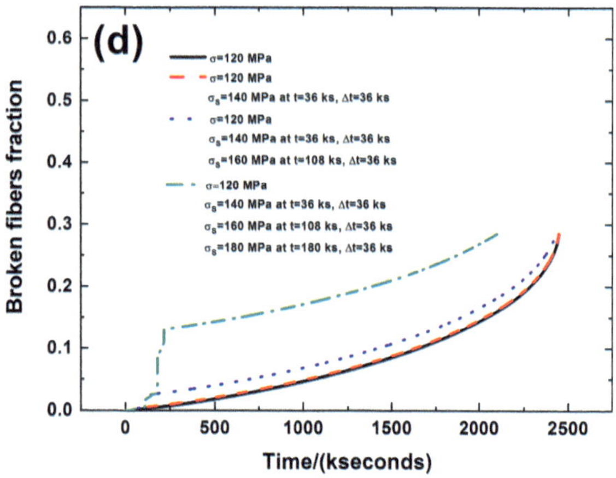

- For the stochastic loading Case 2, the stochastic loading stress of $\sigma_s = 140$ MPa occurs at $t = 36$ ks; the time interval is $\Delta t = 36$ ks; the composite's stress-rupture lifetime is $t = 2446.9$ ks; the time for the composite's interface complete debonding is $t = 36$ ks under the stochastic loading stress of $\sigma_s = 140$ MPa; the time for the composite's interface complete oxidation is $t = 103.5$ ks under constant stress of $\sigma = 120$ MPa; the composite's failure strain is $\varepsilon_c = 0.206\%$; and the corresponding broken fibers fraction is $q = 0.285$.
- For the stochastic loading Case 3, the stochastic loading stresses of $\sigma_s = 140$, 160 MPa occur at $t = 36$, 108 ks; the time interval is $\Delta t = 36$ ks; the composite's stress-rupture lifetime is $t = 2430.8$ ks; the time for the composite's interface complete debonding is $t = 36$ ks under stochastic loading stress of $\sigma_s = 140$ MPa; the time for the composite's interface complete oxidation is $t = 103.5$ ks under constant stress $\sigma = 120$ MPa; the composite's failure strain is $\varepsilon_c = 0.203\%$; and the corresponding broken fibers fraction is $q = 0.285$.
- For the stochastic loading Case 4, the stochastic loading stresses of $\sigma_s = 140$, 160, 180 MPa occur at $t = 36$, 108, 180 ks; the time interval is $\Delta t = 36$ ks; the composite's stress-rupture lifetime is $t = 2099.3$ ks; the time for the composite's interface complete debonding is $t = 36$ ks under stochastic loading stress of $\sigma_s = 140$ MPa; the time for the composite's interface complete oxidation is $t = 103.5$ ks under constant stress $\sigma = 120$ MPa; the composite's failure strain is $\varepsilon_c = 0.191\%$; and the corresponding broken fibers fraction is $q = 0.285$.

When the saturation matrix crack spacing is $l_s = 200$ μm, the damage evolution and lifetime of SiC/SiC composite corresponding to loading spectrums of Cases 1, 2, 3, and 4 at $T = 800$ °C in air atmosphere are analyzed, as follows:

- Under constant stress $\sigma = 120$ MPa, the composite's stress-rupture lifetime is $t = 2447.9$ ks; the time for the composite's interface complete debonding is $t = 181.6$ ks; the time for the composite's interface complete oxidation is $t = 236$ ks;

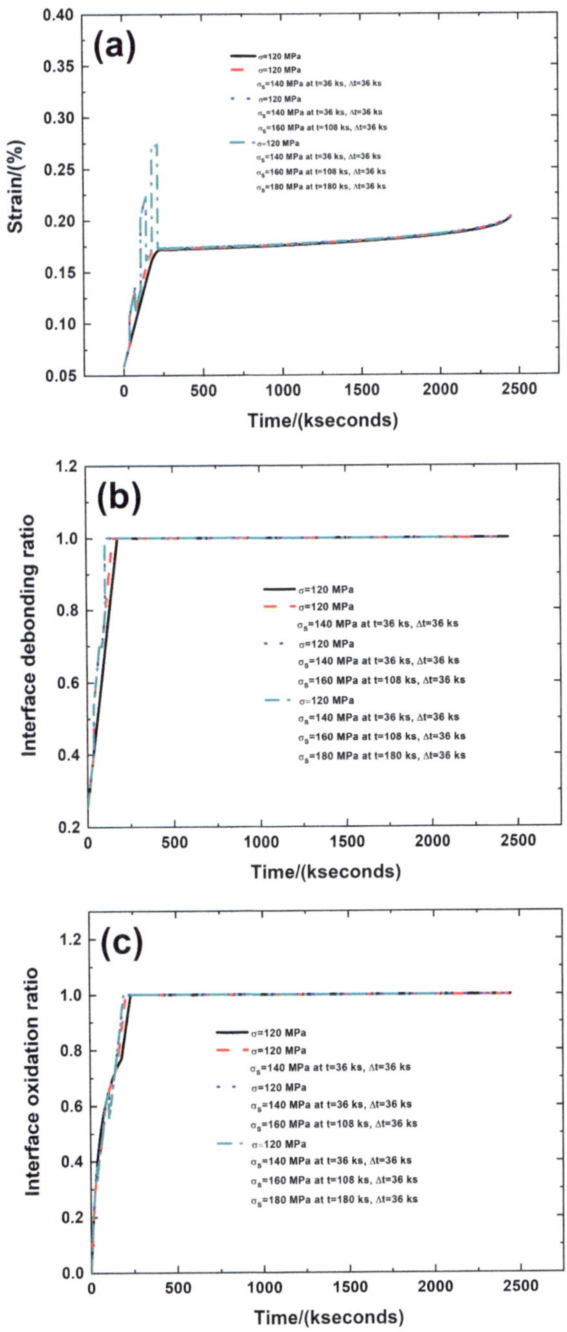

Fig. 3.5 **a** Composite strain versus the time curves; **b** the interface debonding ratio versus the time curves; **c** the interface oxidation ratio versus the time curves; and **d** the broken fibers fraction versus the time curves of SiC/SiC composite under stress-rupture loading of constant stress $\sigma = 120$ MPa, $\sigma_s = 140$, 140/160, 140/160/180 MPa at $t = 36$, 36/108, 36/108/180 ks and $\Delta t = 36$ ks at $T = 800$ °C in air atmosphere when $l_s = 200$ μm

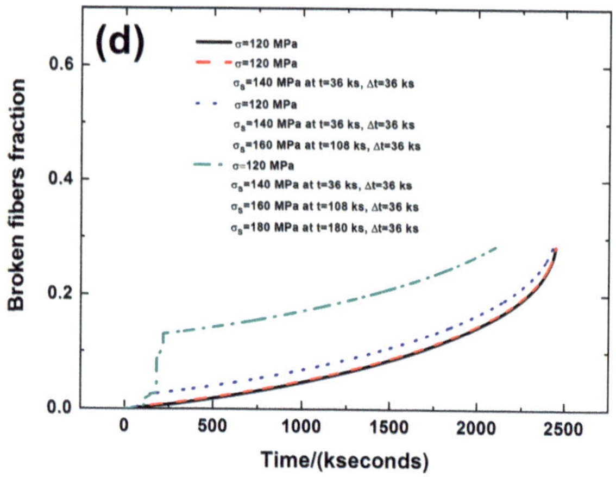

Fig. 3.5 (continued)

the composite's failure strain is $\varepsilon_c = 0.2\%$; and the corresponding broken fibers fraction is $q = 0.285$.

- For the stochastic loading Case 2, the stochastic loading stress of $\sigma_s = 140$ MPa occurs at $t = 36$ ks; the time interval is $\Delta t = 36$ ks; the composite's stress-rupture lifetime is $t = 2446.9$ ks; the time for the composite's interface complete debonding is $t = 152$ ks under the constant stress of $\sigma = 120$ MPa; the time for the composite's interface complete oxidation is $t = 207.3$ ks under constant stress $\sigma = 120$ MPa; the composite's failure strain is $\varepsilon_c = 0.2\%$; and the corresponding broken fibers fraction is $q = 0.285$.
- For the stochastic loading Case 3, the stochastic loading stresses of $\sigma_s = 140$, 160 MPa occur at $t = 36$, 108 ks; the time interval is $\Delta t = 36$ ks; the composite's stress-rupture lifetime is $t = 2430.8$ ks; the time for the composite's interface complete debonding is $t = 109.8$ ks under stochastic loading stress of $\sigma_s = 160$ MPa; the time for the composite's interface complete oxidation is $t = 197.2$ ks under constant stress $\sigma = 120$ MPa; the composite's failure strain is $\varepsilon_c = 0.2\%$; and the broken fibers fraction is $q = 0.285$.
- For the stochastic loading Case 4, the stochastic loading stresses of $\sigma_s = 140$, 160, 180 MPa occur at $t = 36$, 108, 180 ks; the time interval is $\Delta t = 36$ ks; the composite's stress-rupture lifetime is $t = 2099.3$ ks; the time for the composite's interface complete debonding is $t = 109.8$ ks at stochastic loading stress of $\sigma_s = 160$ MPa; the time for the composite's interface complete oxidation is $t = 194.5$ ks under stochastic loading stress of $\sigma = 180$ MPa; the composite's failure strain is $\varepsilon_c = 0.188\%$; and the corresponding broken fibers fraction is $q = 0.285$.

Table 3.3 Composite strain, interface debonding and oxidation ratio, and broken fibers fraction of SiC/SiC composite under stress-rupture loading of constant stress $\sigma = 120$ MPa, $\sigma_s = 140$, 140/160, 140/160/180 MPa at $t = 36$, 36/108, 36/108/180 ks and $\Delta t = 36$ ks at $T = 800$ °C in air atmosphere when $l_s = 100$ μm

σ/MPa	120	120	120	120	120									
t/ks	0	36	59.8	117.8	2447.9									
ε_c/%	0.084	0.126	0.152	0.175	0.206									
η	0.508	0.805	1.0	1.0	1.0									
ω	0	0.38	0.508	1.0	1.0									
q	1×10^{-6}	6.5×10^{-4}	0.0012	0.0028	0.285									
σ/MPa	120	120	140	140	120	120	120							
t/ks	0	36	36	72	72	103.5	2446.9							
ε_c/%	0.084	0.126	0.169	0.197	0.167	0.175	0.206							
η	0.508	0.805	1.0	1.0	1.0	1.0	1.0							
ω	0	0.38	0.348	0.696	0.696	1.0	1.0							
q	1×10^{-6}	6.5×10^{-4}	0.0016	0.0039	0.0039	0.0048	0.285							
σ/MPa	120	120	140	140	120	120	160	160	120	120				
t/ks	0	36	36	72	72	108	108	144	144	2430.8				
ε_c/%	0.084	0.126	0.169	0.197	0.167	0.175	0.236	0.236	0.175	0.203				
η	0.508	0.805	1.0	1.0	1.0	1.0	1.0	1.0	1.0	1.0				
ω	0	0.38	0.348	0.696	0.696	1.0	1.0	1.0	1.0	1.0				
q	1×10^{-6}	6.5×10^{-4}	0.0016	0.0039	0.0039	0.0049	0.017	0.024	0.026	0.285				

(continued)

Table 3.3 (continued)

σ/MPa	120	120	140	140	120	120	160	160	120	120	180	180	120	120
t/ks	0	36	36	72	72	108	108	144	144	180	180	216	216	2099.3
ε_c/%	0.084	0.126	0.169	0.197	0.167	0.175	0.236	0.236	0.175	0.176	0.273	0.276	0.176	0.191
η	0.508	0.805	1.0	1.0	1.0	1.0	1.0	1.0	1.0	1.0	1.0	1.0	1.0	1.0
ω	0	0.38	0.348	0.696	0.696	1.0	1.0	1.0	1.0	1.0	1.0	1.0	1.0	1.0
q	1×10^{-6}	6.5×10^{-4}	0.0016	0.0039	0.0039	0.0049	0.017	0.024	0.026	0.027	0.087	0.11	0.13	0.285

Table 3.4 Composite strain, interface debonding and oxidation ratio, and broken fibers fraction of SiC/SiC composite under stress-rupture loading of constant stress $\sigma = 120$ MPa, $\sigma_s = 140$, 140/160, 140/160/180 MPa at $t = 36$, 36/108, 36/108/180 ks and $\Delta t = 36$ ks at $T = 800$ °C in air atmosphere when $l_s = 200$ μm

σ/MPa	120	120	120	120	120						
t/ks	0	36	181.6	236	2447.9						
ε_c/%	0.058	0.079	0.162	0.172	0.2						
η	0.254	0.402	1.0	1.0	1.0						
ω	0	0.38	0.77	1.0	1.0						
q	1×10^{-6}	6.5×10^{-4}	0.0049	0.0069	0.285						
σ/MPa	120	120	140	140	120	120	120	120			
t/ks	0	36	36	72	72	152	207.3	2446.9			
ε_c/%	0.058	0.079	0.107	0.135	0.112	0.16	0.172	0.2			
η	0.254	0.402	0.529	0.698	0.698	1.0	1.0	1.0			
ω	0	0.38	0.328	0.498	0.498	0.734	1.0	1.0			
q	1×10^{-6}	6.5×10^{-4}	0.0016	0.0039	0.0039	0.0063	0.0082	0.285			
σ/MPa	120	120	140	140	120	120	160	160	120	120	120
t/ks	0	36	36	72	72	108	108	144	144	197.2	2430.8
ε_c/%	0.058	0.079	0.107	0.135	0.112	0.133	0.201	0.223	0.161	0.172	0.2
η	0.254	0.402	0.529	0.698	0.698	0.794	0.991	1.0	1.0	1.0	1.0
ω	0	0.38	0.328	0.498	0.498	0.656	0.553	0.73	0.73	1.0	1.0
q	1×10^{-6}	6.5×10^{-4}	0.0016	0.0039	0.0039	0.0049	0.017	0.024	0.026	0.028	0.285

(continued)

Table 3.4 (continued)

σ/MPa	120	120	140	140	120	120	160	160	120	120	180	180	120	120
t/ks	0	36	36	72	72	108	108	144	144	180	180	216	216	2099.3
ε_c/%	0.058	0.079	0.107	0.135	0.112	0.133	0.201	0.223	0.161	0.171	0.269	0.273	0.173	0.188
η	0.254	0.402	0.529	0.698	0.698	0.794	0.991	1.0	1.0	1.0	1.0	1.0	1.0	1.0
ω	0	0.38	0.328	0.498	0.498	0.656	0.553	0.73	0.73	0.913	0.925	1.0	1.0	1.0
q	1×10^{-6}	6.5×10^{-4}	0.0016	0.0039	0.0039	0.0049	0.017	0.024	0.026	0.027	0.087	0.11	0.13	0.285

3.3.3 Stress-Rupture of SiC/SiC Composite with Stochastic Loading for a Different Interface Debonding Energy

Figures 3.6 and 3.7, and Tables 3.5 and 3.6 show the composite strain, interface debonding and oxidation ratio, and the broken fibers fraction of SiC/SiC composite corresponding to the loading spectrum of Cases 1, 2, 3, and 4 at T = 800 °C in air atmosphere for Γ_d = 0.3 and 0.5 J/m^2. When the interface debonding energy increases, the time for the interface complete debonding increases and the failure strain decreases. The interface debonding length decreases with interface debonding energy, leading to the increase of the time for the complete interface debonding.

When the interface debonding energy is Γ_d = 0.3 J/m^2, the damage evolution and lifetime of SiC/SiC composite corresponding to loading spectrums of Cases 1, 2, 3, and 4 at T = 800 °C in air atmosphere are analyzed, as follows:

- Under constant stress σ = 120 MPa, the composite's stress-rupture lifetime is t = 2447.9 ks; the time for the composite's interface complete debonding is t = 261.9 ks; the time for the composite's interface complete oxidation is t = 295.3 ks; the composite's failure strain is ε_c = 0.201%; and the corresponding broken fibers fraction is q = 0.285.
- For the stochastic loading Case 2, the stochastic loading stress of σ_s = 140 MPa occurs at t = 36 ks; the time interval is Δt = 36 ks; the composite's stress-rupture lifetime is t = 2446.9 ks; the time for the composite's interface complete debonding is t = 224.9 ks under the constant stress of σ = 120 MPa; the time for the composite's interface complete oxidation is t = 259.4 ks under constant stress of σ = 120 MPa; the composite's failure strain is ε_c = 0.202%; and the corresponding broken fibers fraction is q = 0.285.
- For the stochastic loading Case 3, the stochastic loading stresses of σ_s = 140, 160 MPa occur at t = 36, 108 ks; the time interval is Δt = 36 ks; the composite's stress-rupture lifetime is t = 2430.8 ks; the time for the composite's interface complete debonding is t = 211.9 ks under constant stress of σ = 120 MPa; the time for the composite's interface complete oxidation is t = 246.8 ks under constant stress of σ = 120 MPa; the composite's failure strain is ε_c = 0.199%; and the corresponding broken fibers fraction is q = 0.285.
- For the stochastic loading Case 4, the stochastic loading stresses of σ_s = 140, 160, 180 MPa occur at t = 36, 108, 180 ks; the time interval is Δt = 36 ks; the composite's stress-rupture lifetime is t = 2099.3 ks; the time for the composite's interface complete debonding is t = 180 ks under stochastic loading stress of σ_s = 180 MPa; the time for the composite's interface complete oxidation is t = 243.3 ks under constant stress of σ = 120 MPa; the composite's failure strain is ε_c = 0.187%; and the corresponding broken fibers fraction is q = 0.285.

When the interface debonding energy is Γ_d = 0.5 J/m^2, the damage evolution and lifetime of SiC/SiC composite corresponding to loading spectrums of Cases 1, 2, 3, and 4 at T = 800 °C in air atmosphere are analyzed, as follows:

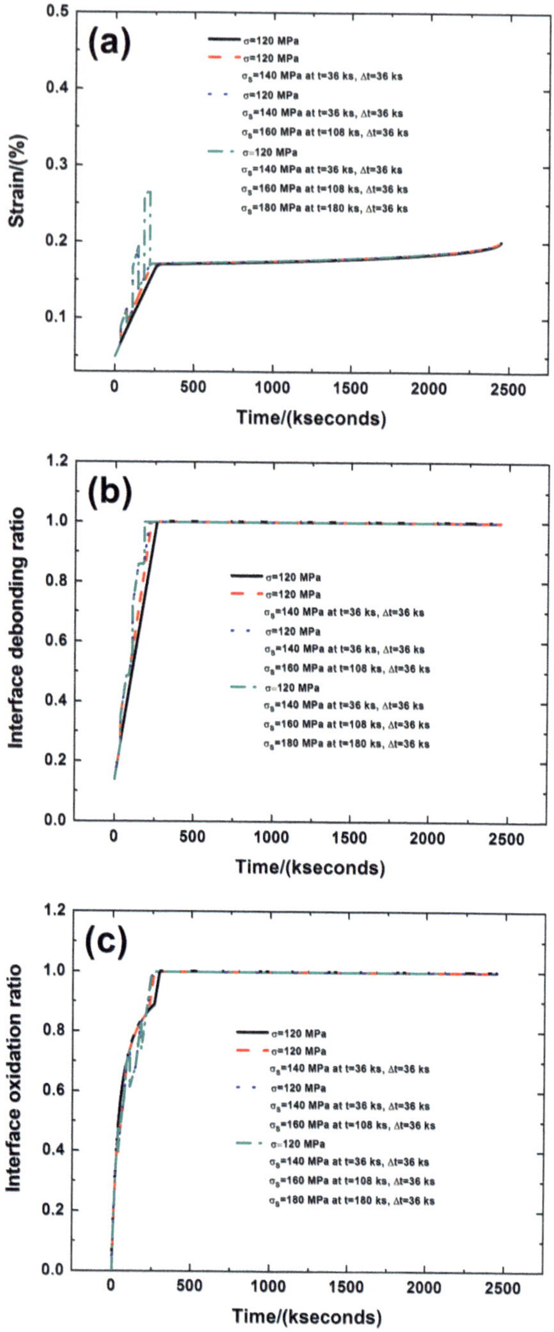

Fig. 3.6 **a** Composite strain versus the time curves; **b** the interface debonding ratio versus the time curves; **c** the interface oxidation ratio versus the time curves; and **d** the broken fibers fraction versus the time curves of SiC/SiC composite under stress-rupture loading of constant stress $\sigma = 120$ MPa, $\sigma_s = 140$, 140/160, 140/160/180 MPa at $t = 36$, 36/108, 36/108/180 ks and $\Delta t = 36$ ks at $T = 800$ °C in air atmosphere when $\Gamma_d = 0.3$ J/m^2

Fig. 3.6 (continued)

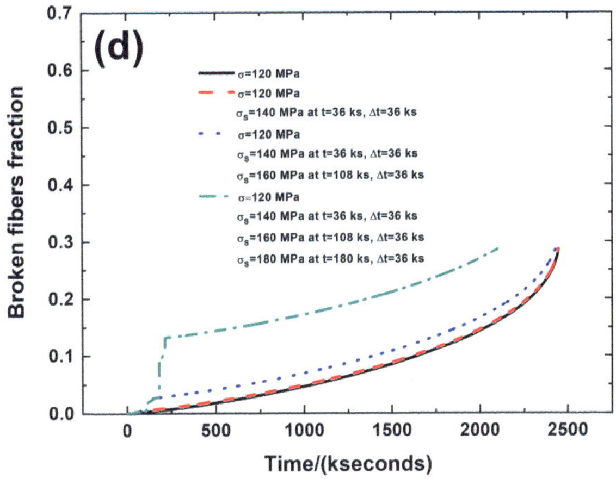

- Under constant stress σ = 120 MPa, the composite's stress-rupture lifetime is t = 2447.9 ks; the time for the composite's interface complete debonding is t = 275.2 ks; the time for the composite's interface complete oxidation is t = 295.3 ks; the composite's failure strain is ε_c = 0.2%; and the corresponding broken fibers fraction is q = 0.285.
- For the stochastic loading Case 2, the stochastic loading stress of σ_s = 140 MPa occurs at t = 36 ks; the time interval is Δt = 36 ks; the composite's stress-rupture lifetime is t = 2446.9 ks; the time for the composite's interface complete debonding is t = 238.1 ks under the constant stress of σ = 120 MPa; the time for the composite's interface complete oxidation is t = 259.4 ks under constant stress of σ = 120 MPa; the composite's failure strain is ε_c = 0.2%; and the corresponding broken fibers fraction is q = 0.285.
- For the stochastic loading Case 3, the stochastic loading stresses of σ_s = 140, 160 MPa occur at t = 36, 108 ks; the time interval is Δt = 36 ks; the composite's stress-rupture lifetime is t = 2430.8 ks; the time for the composite's interface complete debonding is t = 225.1 ks under constant stress of σ = 120 MPa; the time for the composite's interface complete oxidation is t = 246.8 ks under constant stress of σ = 120 MPa; the composite's failure strain is ε_c = 0.199%; and the corresponding broken fibers fraction is q = 0.285.
- For the stochastic loading Case 4, the stochastic loading stresses of σ_s = 140, 160, 180 MPa occur at t = 36, 108, 180 ks; the time interval is Δt = 36 ks; the composite's stress-rupture lifetime is t = 2099.3 ks; the time for the composite's interface complete debonding is t = 180 ks under stochastic loading stress of σ_s = 180 MPa; the time for the composite's interface complete oxidation is t = 243.3 ks under constant stress of σ = 120 MPa; the composite's failure strain is ε_c = 0.187%; and the corresponding broken fibers fraction is q = 0.285.

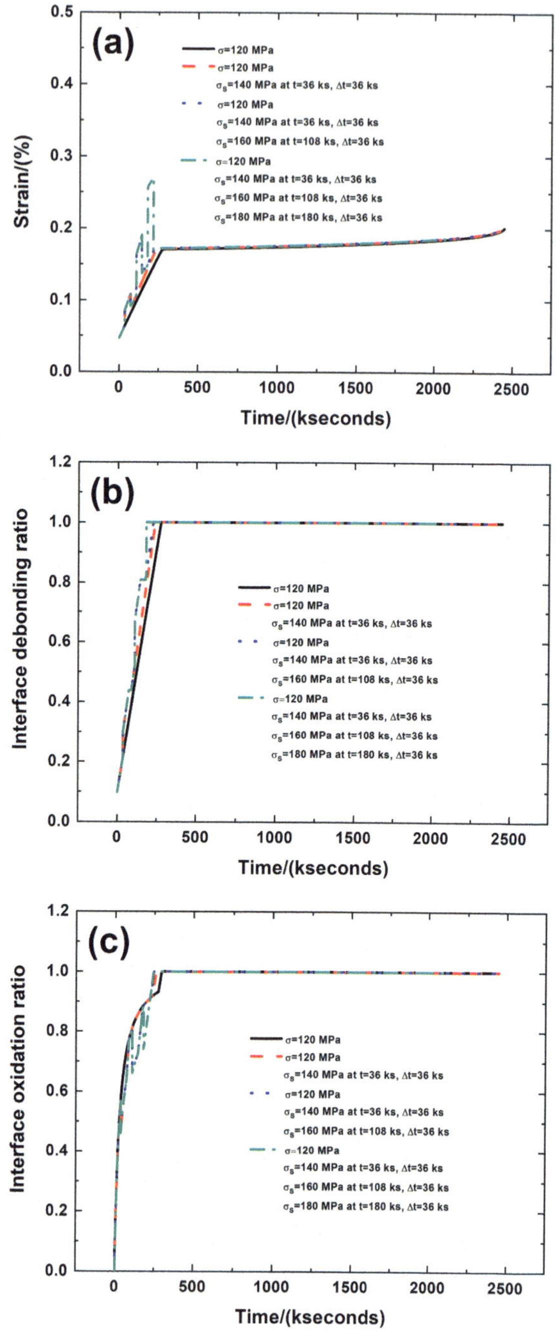

Fig. 3.7 **a** Composite strain versus the time curves; **b** the interface debonding ratio versus the time curves; **c** the interface oxidation ratio versus the time curves; and **d** the broken fibers fraction versus the time curves of SiC/SiC composite under stress-rupture loading of constant stress $\sigma = 120$ MPa, $\sigma_s = 140$, 140/160, 140/160/180 MPa at $t = 36$, 36/108, 36/108/180 ks and $\Delta t = 36$ ks when $\Gamma_d = 0.5$ J/m^2

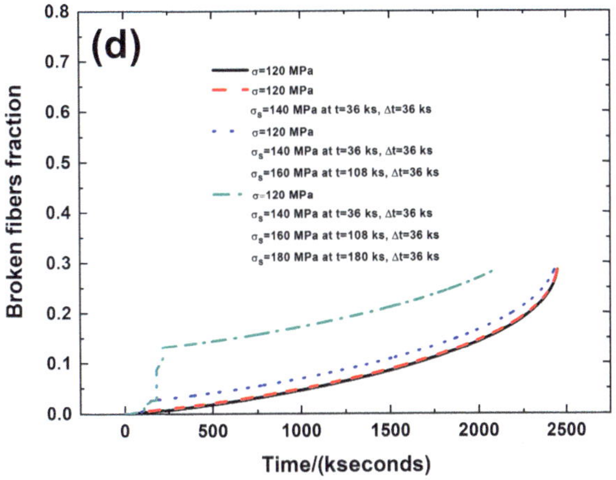

Fig. 3.7 (continued)

3.3.4 Stress-Rupture of SiC/SiC Composite with Stochastic Loading for Different Interface Shear Stress

Figures 3.8 and 3.9, and Tables 3.7 and 3.8 show the composite's strain, interface debonding and oxidation ratio, and the broken fibers fraction of SiC/SiC composite corresponding to the loading spectrum of Cases 1, 2, 3, and 4 at T = 800 °C in air atmosphere for $\tau_i = 20$ and 40 MPa. When the interface shear stress in the slip region increases, the stress-rupture lifetime increases and the time for the interface complete debonding increases. The interface debonding length decreases with the interface shear stress in the slip region, leading to the increase of the time for the interface complete debonding.

When the interface shear stress is $\tau_i = 20$ MPa, the damage evolution and lifetime of SiC/SiC composite corresponding to loading spectrums of Cases 1, 2, 3, and 4 at T = 800 °C in air atmosphere are analyzed, as follows:

- Under constant stress σ = 120 MPa, the composite's stress-rupture lifetime is t = 1769.8 ks; the time for the composite's interface complete debonding is t = 213 ks; the time for the composite's interface complete oxidation is t = 295.3 ks; the composite's failure strain is $\varepsilon_c = 0.2\%$; and the broken fibers fraction is q = 0.285.
- For the stochastic loading Case II, the stochastic loading stress of σ_s = 140 MPa occurs at t = 36 ks; the time interval is Δt = 36 ks; the stress-rupture lifetime is t = 1768.7 ks; the time for the interface complete debonding is t = 175.5 ks under the constant stress of σ = 120 MPa; the time for the composite's interface complete oxidation is t = 259.4 ks under constant stress of σ = 120 MPa; the composite's failure strain is $\varepsilon_c = 0.2\%$; and the corresponding broken fibers fraction is q = 0.285.
- For the stochastic loading Case 3, the stochastic loading stresses of σ_s = 140, 160 MPa occur at t = 36, 108 ks; the time interval is Δt = 36 ks; the composite's

Table 3.5 Composite strain, interface debonding and oxidation ratio, and broken fibers fraction of SiC/SiC composite under stress-rupture loading of constant stress $\sigma = 120$ MPa, $\sigma_s = 140$, 140/160, 140/160/180 MPa at $t = 36$, 36/108, 36/108/180 ks and $\Delta t = 36$ ks at $T = 800$ °C in air atmosphere when $\Gamma_d = 0.3$ J/m^2

σ/MPa	120	120	120	120	120										
t/ks	0	36	261.9	295.3	2447.9										
ε_c/%	0.049	0.067	0.167	0.17	0.201										
η	0.14	0.259	1.0	1.0	1.0										
ω	0	0.472	0.887	1.0	1.0										
q	1×10^{-6}	6.5×10^{-4}	0.0078	0.009	0.285										
σ/MPa	120	120	140	140	120	120	120	120							
t/ks	0	36	36	72	72	224.9	259.4	2446.9							
ε_c/%	0.049	0.067	0.089	0.112	0.094	0.167	0.171	0.202							
η	0.14	0.259	0.352	0.487	0.487	1.0	1.0	1.0							
ω	0	0.472	0.395	0.571	0.571	0.867	1.0	1.0							
q	1×10^{-6}	6.5×10^{-4}	0.0016	0.0039	0.0039	0.0088	0.01	0.285							
σ/MPa	120	120	140	140	120	120	160	160	120	120	120	120			
t/ks	0	36	36	72	72	108	108	144	144	211.9	246.8	2430.8			
ε_c/%	0.049	0.067	0.089	0.112	0.094	0.109	0.167	0.194	0.138	0.167	0.171	0.199			
η	0.14	0.259	0.352	0.487	0.487	0.564	0.716	0.86	0.86	1.0	1.0	1.0			
ω	0	0.472	0.395	0.571	0.571	0.74	0.612	0.679	0.679	0.859	1.0	1.0			
q	1×10^{-6}	6.5×10^{-4}	0.0016	0.0039	0.0039	0.0049	0.017	0.024	0.026	0.028	0.03	0.285			

(continued)

Table 3.5 (continued)

σ/MPa	120	120	140	140	120	120	160	160	120	120	180	180	120	120	120
t/ks	0	36	36	72	72	108	108	144	144	180	180	216	216	243.3	2099.3
ε_c/%	0.049	0.067	0.089	0.112	0.094	0.109	0.167	0.194	0.138	0.152	0.256	0.269	0.169	0.171	0.187
η	0.14	0.259	0.352	0.487	0.487	0.564	0.716	0.86	0.86	0.875	1.0	1.0	1.0	1.0	1.0
ω	0	0.472	0.395	0.571	0.571	0.74	0.612	0.679	0.679	0.834	0.74	0.888	0.888	1.0	1.0
q	1×10^{-6}	6.5×10^{-4}	0.0016	0.0039	0.0039	0.0049	0.017	0.024	0.026	0.027	0.087	0.11	0.13	0.133	0.285

Table 3.6 Composite strain, interface debonding and oxidation ratio, and broken fibers fraction of SiC/SiC composite under stress-rupture loading of constant stress $\sigma = 120$ MPa, $\sigma_s = 140$, 140/160, 140/160/180 MPa at $t = 36$, 36/108, 36/108/180 ks and $\Delta t = 36$ ks at $T = 800$ °C in air atmosphere when $\Gamma_d = 0.5$ J/m^2

σ/MPa	120	120	120	120	120										
t/ks	0	36	275.2	295.3	2447.9										
ε_c/%	0.046	0.063	0.169	0.17	0.2										
η	0.097	0.215	1.0	1.0	1.0										
ω	0	0.567	0.932	1.0	1.0										
q	1×10^{-6}	6.5×10^{-4}	0.008	0.009	0.285										
σ/MPa	120	120	140	140	120	120	120	120							
t/ks	0	36	36	72	72	238.1	259.4	2446.9							
ε_c/%	0.046	0.063	0.085	0.108	0.091	0.169	0.171	0.2							
η	0.097	0.215	0.302	0.438	0.438	1.0	1.0	1.0							
ω	0	0.567	0.46	0.636	0.636	0.918	1.0	1.0							
q	1×10^{-6}	6.5×10^{-4}	0.0016	0.0039	0.0039	0.029	0.01	0.285							
σ/MPa	120	120	140	140	120	120	160	160	120	120	120	120			
t/ks	0	36	36	72	72	108	108	144	144	225.1	246.8	2430.8			
ε_c/%	0.046	0.063	0.085	0.108	0.091	0.106	0.162	0.19	0.137	0.17	0.171	0.199			
η	0.097	0.215	0.302	0.438	0.438	0.514	0.665	0.808	0.808	1.0	1.0	1.0			
ω	0	0.567	0.46	0.636	0.636	0.81	0.659	0.723	0.723	0.912	1.0	1.0			
q	1×10^{-6}	6.5×10^{-4}	0.0016	0.0039	0.0039	0.0049	0.017	0.024	0.026	0.132	0.03	0.285			

(continued)

Table 3.6 (continued)

σ/MPa	120	120	140	140	120	120	160	160	120	120	180	180	120	120	120
t/ks	0	36	36	72	72	108	108	144	144	180	180	216	216	243.3	2099.3
ε_c/%	0.046	0.063	0.085	0.108	0.091	0.106	0.162	0.19	0.137	0.148	0.256	0.269	0.169	0.171	0.187
η	0.097	0.215	0.302	0.438	0.438	0.514	0.665	0.808	0.808	0.823	1.0	1.0	1.0	1.0	1.0
ω	0	0.567	0.46	0.636	0.636	0.81	0.659	0.723	0.723	0.886	0.74	0.888	0.888	1.0	1.0
q	1×10^{-6}	6.5×10^{-4}	0.0016	0.0039	0.0039	0.0049	0.017	0.024	0.026	0.027	0.087	0.11	0.13	0.133	0.285

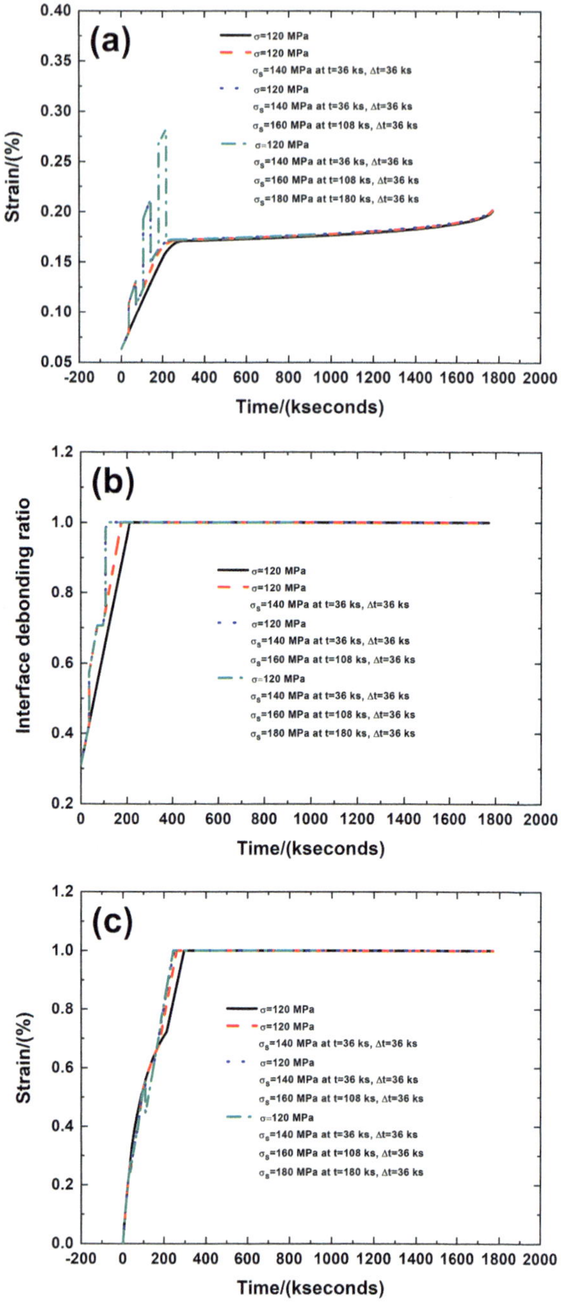

Fig. 3.8 **a** Composite strain versus the time curves; **b** the interface debonding ratio versus the time curves; **c** the interface oxidation ratio versus the time curves; and **d** the broken fibers fraction versus time curves of SiC/SiC composite under stress-rupture loading of constant stress $\sigma = 120$ MPa, $\sigma_s = 140$, 140/160, 140/160/180 MPa at $t = 36$, 36/108, 36/108/180 ks and $\Delta t = 36$ ks at $T = 800$ °C in air atmosphere when $\tau_i = 20$ MPa

Fig. 3.8 (continued)

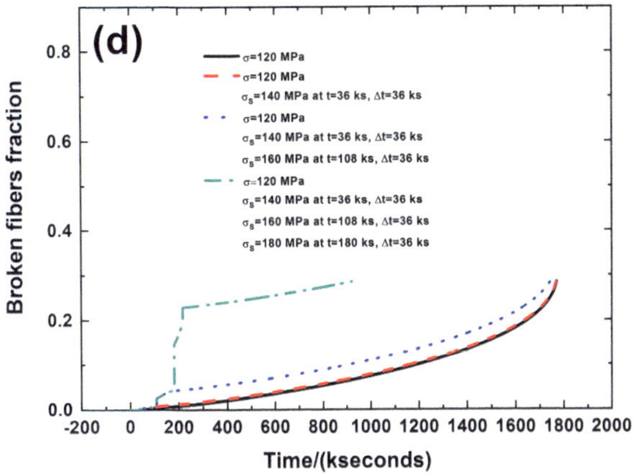

stress-rupture lifetime is t = 1745.3 ks; the time for the composite's interface complete debonding is t = 112.7 ks under stochastic loading stress of σ_s = 160 MPa; the time for the composite's interface complete oxidation is t = 246.8 ks under constant stress of σ = 120 MPa; the composite's failure strain is ε_c = 0.198%; and the corresponding broken fibers fraction is q = 0.285.

- For the stochastic loading Case 4, the stochastic loading stresses of σ_s = 140, 160, 180 MPa occur at t = 36, 108, 180 ks; the time interval is Δt = 36 ks; the composite's stress-rupture lifetime is t = 921.6 ks; the time for the composite's interface complete debonding is t = 112.7 ks under stochastic loading stress of σ_s = 160 MPa; the time for the composite's interface complete oxidation is t = 243.3 ks under constant stress of σ = 120 MPa; the composite's failure strain is ε_c = 0.177%; and the composite's broken fibers fraction is q = 0.285.

When the interface shear stress is τ_i = 40 MPa, the damage evolution and lifetime of SiC/SiC composite corresponding to loading spectrums of Cases 1, 2, 3, and 4 at T = 800 °C in air atmosphere are analyzed, as follows:

- Under constant stress σ = 120 MPa, the composite's stress-rupture lifetime is t = 3081.4 ks; the time for the composite's interface complete debonding is t = 257.1 ks; the time for the composite's interface complete oxidation is t = 295.3 ks; the composite's failure strain is ε_c = 0.2%; and the corresponding broken fibers fraction is q = 0.285.
- For the stochastic loading Case 2, the stochastic loading stress of σ_s = 140 MPa occurs at t = 36 ks; the time interval is Δt = 36 ks; the composite's stress-rupture lifetime is t = 3080.5 ks; the time for the composite's interface complete debonding is t = 220.4 ks under the constant stress of σ = 120 MPa; the time for the composite's interface complete oxidation is t = 259.4 ks under constant stress of σ = 120 MPa; the composite's failure strain is ε_c = 0.2%; and the corresponding broken fibers fraction is q = 0.285.
- For the stochastic loading Case 3, the stochastic loading stresses of σ_s = 140, 160 MPa occur at t = 36, 108 ks; the time interval is Δt = 36 ks; the composite's

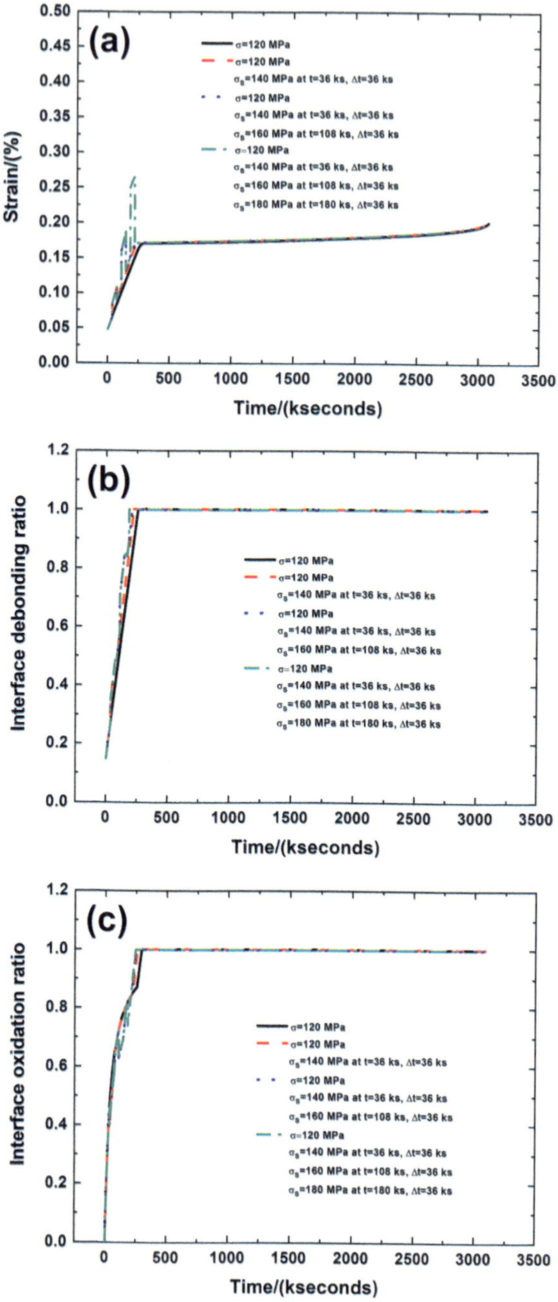

Fig. 3.9 **a** Composite strain versus the time curves; **b** the interface debonding ratio versus the time curves; **c** the interface oxidation ratio versus the time curves; and **d** the broken fibers fraction versus the time curves of SiC/SiC composite under stress-rupture loading of constant stress $\sigma = 120$ MPa, $\sigma_s = 140$, 140/160, 140/160/180 MPa at $t = 36$, 36/108, 36/108/180 ks and $\Delta t = 36$ ks at $T = 800$ °C in air atmosphere when $\tau_i =$ 40 MPa

Fig. 3.9 (continued)

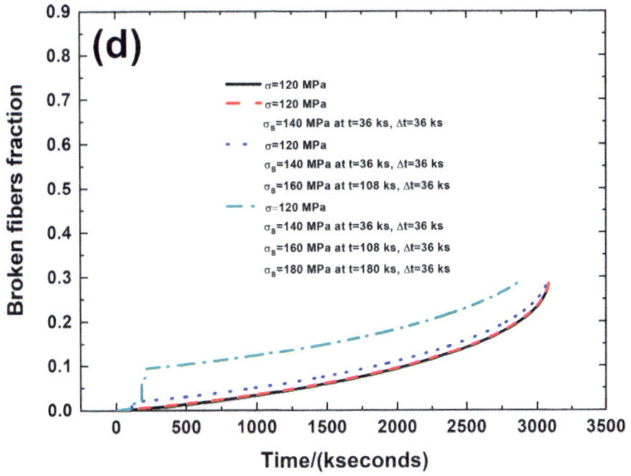

stress-rupture lifetime is $t = 3067.6$ ks; the time for the composite's interface complete debonding is $t = 207.5$ ks under constant stress of $\sigma = 120$ MPa; the time for the composite's interface complete oxidation is $t = 246.8$ ks under constant stress of $\sigma = 120$ MPa; the composite's failure strain is $\varepsilon_c = 0.2\%$; and the corresponding broken fibers fraction is $q = 0.285$.

- For the stochastic loading Case 4, the stochastic loading stresses of $\sigma_s = 140$, 160, 180 MPa occur at $t = 36$, 108, 180 ks; the time interval is $\Delta t = 36$ ks; the composite's stress-rupture lifetime is $t = 2857.3$ ks; the time for the composite's interface complete debonding is $t = 180$ ks under stochastic loading stress of $\sigma_s = 180$ MPa; the time for the composite's interface complete oxidation is $t = 243.3$ ks under constant stress of $\sigma = 120$ MPa; the composite's failure strain is $\varepsilon_c = 0.191\%$; and the corresponding broken fibers fraction is $q = 0.285$.

Figures 3.10 and 3.11, and Tables 3.9 and 3.10 show the composite strain, interface debonding and oxidation ratio, and the broken fibers fraction of SiC/SiC composite corresponding to the loading sequences of Cases 1, 2, 3, and 4 at $T = 800$ °C in air atmosphere for $\tau_f = 2$ and 4 MPa. When the interface shear stress in the oxidation region increases, the time for the interface complete debonding increases and the failure strain decreases. The interface debonding length decreases with the interface shear stress in the oxidation region, leading to the increase of the time for the interface complete debonding.

When the interface shear stress is $\tau_f = 2$ MPa, the damage evolution and lifetime of SiC/SiC composite corresponding to loading spectrums of Cases 1, 2, 3, and 4 at $T = 800$ °C in air atmosphere are analyzed, as follows:

- Under constant stress $\sigma = 120$ MPa, the composite's stress-rupture lifetime is $t = 2447.9$ ks; the time for the composite's interface complete debonding is $t = 251.3$ ks; the time for the composite's interface complete oxidation is $t = 295.3$ ks; the composite's failure strain is $\varepsilon_c = 0.193\%$; and the corresponding broken fibers fraction is $q = 0.285$.

Table 3.7 Composite strain, fiber/matrix interface debonding and oxidation ratio, and broken fibers fraction of SiC/SiC composite under stress-rupture loading of constant stress $\sigma = 120$ MPa, $\sigma_s = 140$, 140/160, 140/160/180 MPa at $t = 36$, 36/108, 36/108/180 ks and $\Delta t = 36$ ks at $T = 800$ °C in air atmosphere when $\tau_i = 20$ MPa

σ/MPa	120	120	120	120	120										
t/ks	0	36	213	295.3	1769.8										
ε_c/%	0.063	0.08	0.158	0.171	0.2										
η	0.31	0.427	1.0	1.0	1.0										
ω	0	0.286	0.722	1.0	1.0										
q	2×10^{-6}	9.7×10^{-4}	0.009	0.014	0.285										
σ/MPa	120	120	140	140	120	120	120	120							
t/ks	0	36	36	72	72	175.5	259.4	1768.7							
ε_c/%	0.063	0.08	0.108	0.131	0.107	0.157	0.171	0.2							
η	0.31	0.427	0.572	0.706	0.706	1.0	1.0	1.0							
ω	0	0.286	0.243	0.394	0.394	0.677	1.0	1.0							
q	2×10^{-6}	9.7×10^{-4}	0.0024	0.0059	0.0059	0.01	0.015	0.285							
σ/MPa	120	120	140	140	120	120	160	160	120	120	120				
t/ks	0	36	36	72	72	108	108	144	144	246.8	1745.3				
ε_c/%	0.063	0.08	0.108	0.131	0.107	0.124	0.192	0.212	0.149	0.172	0.198				
η	0.31	0.427	0.572	0.706	0.706	0.751	0.981	1.0	1.0	1.0	1.0				
ω	0	0.286	0.243	0.394	0.394	0.555	0.447	0.584	0.584	1.0	1.0				
q	2×10^{-6}	9.7×10^{-4}	0.0024	0.0059	0.0059	0.007	0.026	0.037	0.04	0.046	0.285				

(continued)

Table 3.7 (continued)

σ/MPa	120	120	140	140	120	120	160	160	120	120	180	180	120	120	120
t/ks	0	36	36	72	72	108	108	144	144	180	180	216	216	243.3	921.6
ε_c/%	0.063	0.08	0.108	0.131	0.107	0.124	0.192	0.212	0.149	0.162	0.267	0.282	0.17	0.172	0.177
η	0.31	0.427	0.572	0.706	0.706	0.751	0.981	1.0	1.0	1.0	1.0	1.0	1.0	1.0	1.0
ω	0	0.286	0.243	0.394	0.394	0.555	0.447	0.584	0.584	0.73	0.74	0.888	0.888	1.0	1.0
q	2×10^{-6}	9.7×10^{-4}	0.0024	0.0059	0.0059	0.007	0.026	0.037	0.04	0.042	0.145	0.193	0.228	0.23	0.285

Table 3.8 Composite strain, fiber/matrix interface debonding and oxidation ratio, and broken fibers fraction of SiC/SiC composite under stress-rupture loading of constant stress $\sigma = 120$ MPa, $\sigma_s = 140$, 140/160, 140/160/180 MPa at $t = 36$, 36/108, 36/108/180 ks and $\Delta t = 36$ ks at $T = 800$ °C in air atmosphere when $\tau_i = 40$ MPa

σ/MPa	120	120	120	120	120										
t/ks	0	36	257.1	295.3	3081.4										
ε_c/%	0.048	0.065	0.165	0.17	0.2										
η	0.149	0.269	1.0	1.0	1.0										
ω	0	0.455	0.871	1.0	1.0										
q	1×10^{-6}	4.8×10^{-4}	0.0057	0.0068	0.285										
σ/MPa	120	120	140	140	120	120	120	120							
t/ks	0	36	36	72	72	220.4	259.4	3080.5							
ε_c/%	0.048	0.065	0.085	0.108	0.09	0.164	0.171	0.2							
η	0.149	0.269	0.349	0.485	0.485	1.0	1.0	1.0							
ω	0	0.455	0.399	0.574	0.574	0.85	1.0	1.0							
q	1×10^{-6}	4.8×10^{-4}	0.0012	0.0029	0.0029	0.006	0.0076	0.285							
σ/MPa	120	120	140	140	120	120	160	160	120	120	120	120			
t/ks	0	36	36	72	72	108	108	144	144	207.5	246.8	3067.6			
ε_c/%	0.048	0.065	0.085	0.108	0.09	0.108	0.159	0.187	0.132	0.164	0.171	0.2			
η	0.149	0.269	0.349	0.485	0.485	0.577	0.699	0.843	0.843	1.0	1.0	1.0			
ω	0	0.455	0.399	0.574	0.574	0.722	0.627	0.693	0.693	0.84	1.0	1.0			
q	1×10^{-6}	4.8×10^{-4}	0.0012	0.0029	0.0029	0.0037	0.012	0.017	0.019	0.021	0.022	0.285			

(continued)

Table 3.8 (continued)

σ/MPa	120	120	140	140	120	120	160	160	120	120	180	180	120	120	120
t/ks	0	36	36	72	72	108	108	144	144	180	180	216	216	243.3	2857.3
ε_c/%	0.048	0.065	0.085	0.108	0.09	0.108	0.159	0.187	0.132	0.151	0.249	0.265	0.168	0.171	0.191
η	0.149	0.269	0.349	0.485	0.485	0.577	0.699	0.843	0.843	0.891	1.0	1.0	1.0	1.0	1.0
ω	0	0.455	0.399	0.574	0.574	0.722	0.627	0.693	0.693	0.819	0.74	0.888	0.888	1.0	1.0
q	1×10^{-6}	4.8×10^{-4}	0.0012	0.0029	0.0029	0.0037	0.012	0.017	0.019	0.02	0.063	0.077	0.094	0.095	0.285

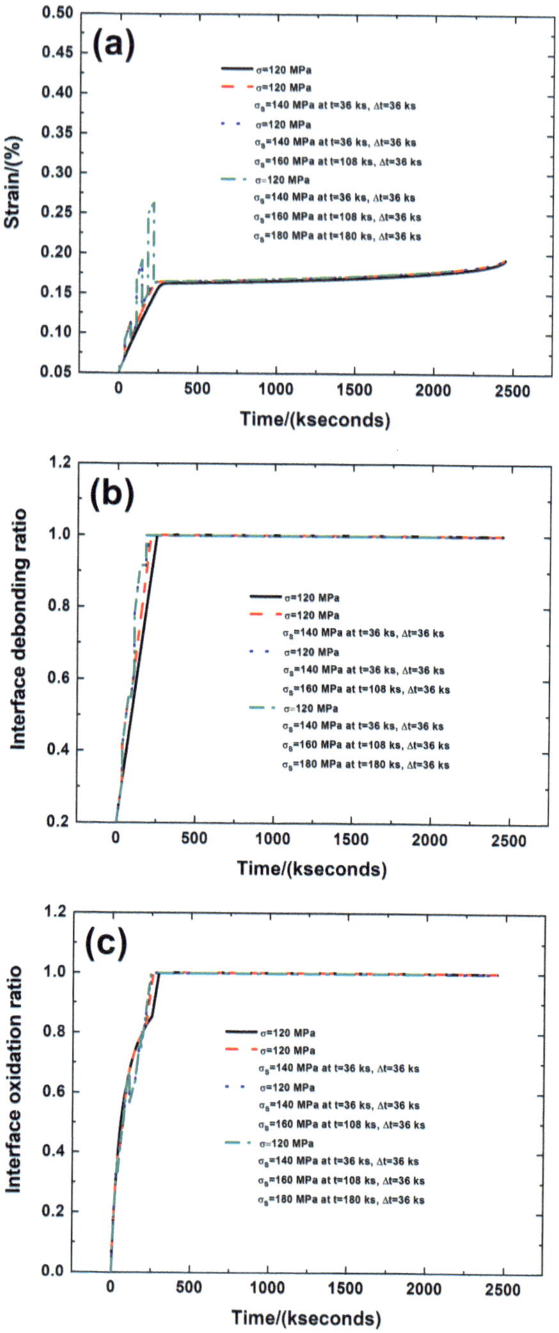

Fig. 3.10 **a** Composite strain versus the time curves; **b** the fiber/matrix interface debonding ratio versus the time curves; **c** the interface oxidation ratio versus the time curves; and **d** the broken fibers fraction versus the time curves of SiC/SiC composite under stress-rupture loading of constant stress $\sigma = 120$ MPa, $\sigma_s = 140$, 140/160, 140/160/180 MPa at $t = 36$, 36/108, 36/108/180 ks and $\Delta t = 36$ ks at $T = 800$ °C in air atmosphere when $\tau_f = 2$ MPa

Fig. 3.10 (continued)

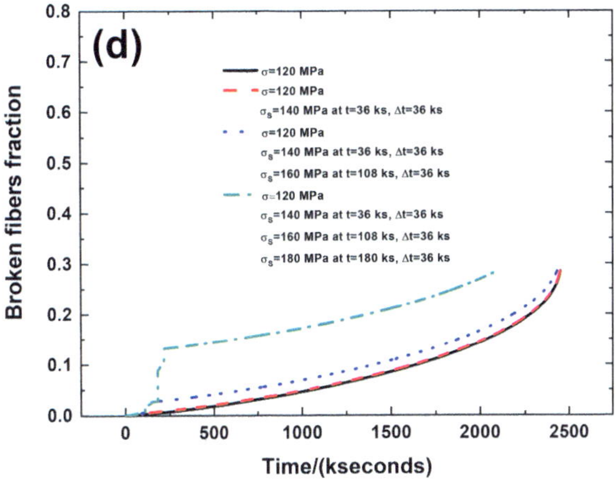

- For the stochastic loading Case 2, the stochastic loading stress of $\sigma_s = 140$ MPa occurs at $t = 36$ ks; the time spacing is $\Delta t = 36$ ks; the composite's stress-rupture lifetime is $t = 2446.9$ ks; the time for the composite's interface complete debonding is $t = 213$ ks under the constant stress of $\sigma = 120$ MPa; the time for the composite's interface complete oxidation is $t = 259.4$ ks under constant stress of $\sigma = 120$ MPa; the composite's failure strain is $\varepsilon_c = 0.194\%$; and the corresponding broken fibers fraction is $q = 0.285$.
- For the stochastic loading Case 3, the stochastic loading stresses of $\sigma_s = 140$, 160 MPa occur at $t = 36$, 108 ks; the time interval is $\Delta t = 36$ ks; the composite's stress-rupture lifetime is $t = 2430.8$ ks; the time for the composite's interface complete debonding is $t = 199.5$ ks under constant stress of $\sigma = 120$ MPa; the time for the composite's interface complete oxidation is $t = 246.8$ ks under constant stress of $\sigma = 120$ MPa; the composite's failure strain is $\varepsilon_c = 0.192\%$; and the corresponding broken fibers fraction is $q = 0.285$.
- For the stochastic loading Case 4, the stochastic loading stresses of $\sigma_s = 140$, 160, 180 MPa occur at $t = 36$, 108, 180 ks; the time interval is $\Delta t = 36$ ks; the composite's stress-rupture lifetime is $t = 2099.3$ ks; the time for the composite's interface complete debonding is $t = 180$ ks under stochastic loading stress of $\sigma_s = 180$ MPa; the time for the composite's interface complete oxidation is $t = 243.3$ ks under constant stress of $\sigma = 120$ MPa; the composite's failure strain is $\varepsilon_c = 0.18\%$; and the corresponding broken fibers fraction is $q = 0.285$.

When the fiber/matrix interface shear stress is $\tau_f = 4$ MPa, the damage evolution and lifetime of SiC/SiC composite corresponding to loading spectrums of Cases 1, 2, 3, and 4 at $T = 800$ °C in air atmosphere are analyzed, as follows:

- Under constant stress $\sigma = 120$ MPa, the composite's stress-rupture lifetime is $t = 2447.9$ ks; the time for the composite's interface complete debonding is $t = 270.7$ ks; the time for the composite's interface complete oxidation is $t = 295.3$ ks;

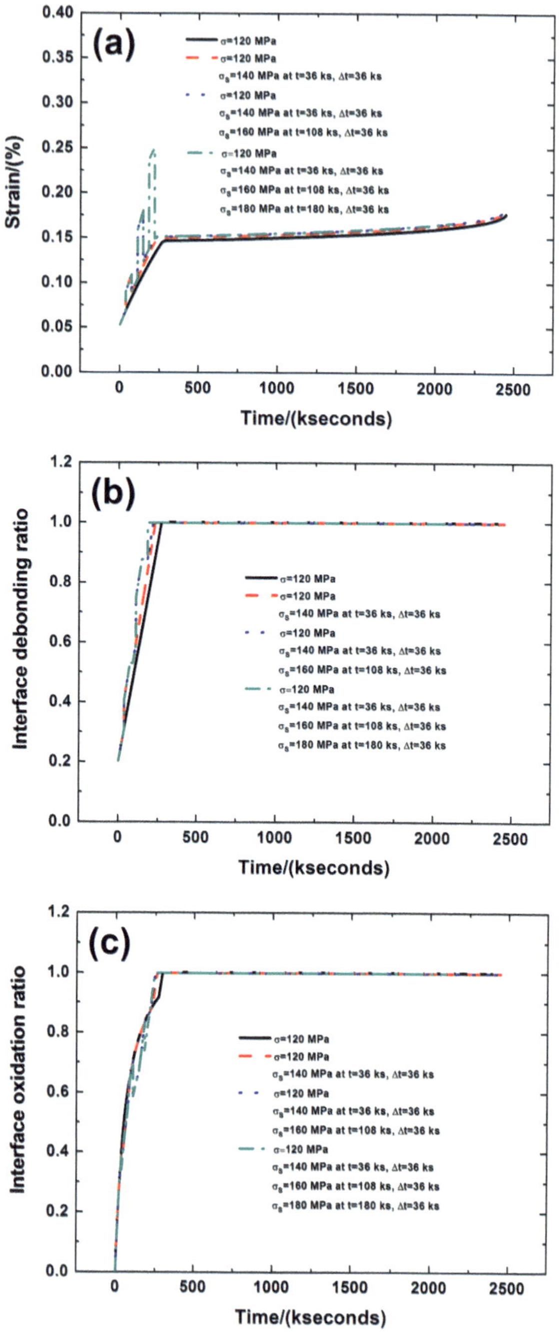

Fig. 3.11 **a** Composite strain versus the time curves; **b** the interface debonding ratio versus the time curves; **c** the interface oxidation ratio versus the time curves; and **d** the broken fibers fraction versus the time curves of SiC/SiC composite under stress-rupture loading of constant stress $\sigma = 120$ MPa, $\sigma_s = 140$, 140/160, 140/160/180 MPa at $t = 36$, 36/108, 36/108/180 ks and $\Delta t = 36$ ks at $T = 800$ °C in air atmosphere when $\tau_f = 4$ MPa

Fig. 3.11 (continued)

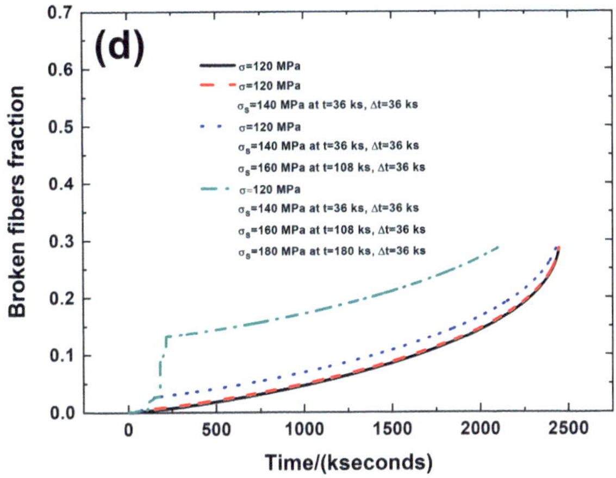

the composite's failure strain is $\varepsilon_c = 0.177\%$; and the composite's broken fibers fraction is $q = 0.285$.

- For the stochastic loading Case 2, the stochastic loading stress of $\sigma_s = 140$ MPa occurs at $t = 36$ ks; the time interval is $\Delta t = 36$ ks; the composite's stress-rupture lifetime is $t = 2446.9$ ks; the time for the composite's interface complete debonding is $t = 229.4$ ks under the constant stress of $\sigma = 120$ MPa; the time for the composite's interface complete oxidation is $t = 259.4$ ks under constant stress of $\sigma = 120$ MPa; the composite's failure strain is $\varepsilon_c = 0.18\%$; and the corresponding broken fibers fraction is $q = 0.285$.
- For the stochastic loading Case 3, the stochastic loading stresses of $\sigma_s = 140$, 160 MPa occur at $t = 36$, 108 ks; the time interval is $\Delta t = 36$ ks; the composite's stress-rupture lifetime is $t = 2430.8$ ks; the time for the composite's interface complete debonding is $t = 214.9$ ks under constant stress of $\sigma = 120$ MPa; the time for the composite's interface complete oxidation is $t = 246.8$ ks under constant stress of $\sigma = 120$ MPa; the composite's failure strain is $\varepsilon_c = 0.179\%$; and the corresponding broken fibers fraction is $q = 0.285$.
- For the stochastic loading Case 4, the stochastic loading stresses of $\sigma_s = 140$, 160, 180 MPa occur at $t = 36$, 108, 180 ks; the time interval is $\Delta t = 36$ ks; the composite's stress-rupture lifetime is $t = 2099.3$ ks; the time for the composite's interface complete debonding is $t = 180$ ks under stochastic loading stress of $\sigma_s = 180$ MPa; the time for the composite's interface complete oxidation is $t = 243.3$ ks under constant stress of $\sigma = 120$ MPa; the composite's failure strain is $\varepsilon_c = 0.167\%$; and the corresponding broken fibers fraction is $q = 0.285$.

Table 3.9 Composite strain, fiber/matrix interface debonding and oxidation ratio, and broken fibers fraction of SiC/SiC composite under stress-rupture loading of constant stress $\sigma = 120$ MPa, $\sigma_S = 140$, 140/160, 140/160/180 MPa at $t = 36$, 36/108, 36/108/180 ks and $\Delta t = 36$ ks at $T = 800$ °C in air atmosphere when $\tau_f = 2$ MPa

σ/MPa	120	120	120	120	120										
t/ks	0	36	251.3	295.3	2447.9										
ε_c/%	0.053	0.069	0.157	0.162	0.193										
η	0.203	0.318	1.0	1.0	1.0										
ω	0	0.385	0.851	1.0	1.0										
q	1×10^{-6}	6.5×10^{-4}	0.007	0.009	0.285										
σ/MPa	120	120	140	140	120	120	120	120							
t/ks	0	36	36	72	72	213	259.4	2446.9							
ε_c/%	0.053	0.069	0.092	0.113	0.094	0.158	0.164	0.194							
η	0.203	0.318	0.419	0.549	0.549	1.0	1.0	1.0							
ω	0	0.385	0.332	0.506	0.506	0.822	1.0	1.0							
q	1×10^{-6}	6.5×10^{-4}	0.0016	0.0039	0.0039	0.008	0.01	0.285							
σ/MPa	120	120	140	140	120	120	160	160	120	120	120	120			
t/ks	0	36	36	72	72	108	108	144	144	199.5	246.8	2430.8			
ε_c/%	0.053	0.069	0.092	0.113	0.094	0.11	0.166	0.192	0.133	0.158	0.165	0.192			
η	0.203	0.318	0.419	0.549	0.549	0.621	0.778	0.917	0.917	1.0	1.0	1.0			
ω	0	0.385	0.332	0.506	0.506	0.671	0.563	0.637	0.637	0.809	1.0	1.0			
q	1×10^{-6}	6.5×10^{-4}	0.0016	0.0039	0.0039	0.0049	0.017	0.024	0.026	0.028	0.03	0.285			

(continued)

Table 3.9 (continued)

σ/MPa	120	120	140	140	120	120	160	160	120	120	180	180	120	120	120
t/ks	0	36	36	72	72	108	108	144	144	180	180	216	216	243.3	2099.3
ε_c/%	0.053	0.069	0.092	0.113	0.094	0.11	0.166	0.192	0.133	0.149	0.249	0.263	0.162	0.165	0.18
η	0.203	0.318	0.419	0.549	0.549	0.621	0.778	0.917	0.917	0.926	1.0	1.0	1.0	1.0	1.0
ω	0	0.385	0.332	0.506	0.506	0.671	0.563	0.637	0.637	0.788	0.74	0.888	0.888	1.0	1.0
q	1×10^{-6}	6.5×10^{-4}	0.0016	0.0039	0.0039	0.0049	0.017	0.024	0.026	0.027	0.087	0.109	0.132	0.133	0.285

Table 3.10 Composite strain, interface debonding and oxidation ratio, and broken fibers fraction of SiC/SiC composite under stress-rupture loading of constant stress $\sigma = 120$ MPa, $\sigma_s = 140$, 140/160, 140/160/180 MPa at $t = 36$, 36/108, 36/108/180 ks and $\Delta t = 36$ ks at $T = 800$ °C in air atmosphere when $\tau_f = 4$ MPa

σ/MPa	120	120	120	120	120										
t/ks	0	36	270.7	295.3	2447.9										
ε_c/%	0.053	0.068	0.145	0.146	0.177										
η	0.203	0.309	1.0	1.0	1.0										
ω	0	0.395	0.917	1.0	1.0										
q	1×10^{-6}	6.5×10^{-4}	0.008	0.009	0.285										
σ/MPa	120	120	140	140	120	120	120	120							
t/ks	0	36	36	72	72	229.4	259.4	2446.9							
ε_c/%	0.053	0.068	0.09	0.109	0.09	0.147	0.15	0.18							
η	0.203	0.309	0.409	0.531	0.531	1.0	1.0	1.0							
ω	0	0.395	0.34	0.524	0.524	0.885	1.0	1.0							
q	1×10^{-6}	6.5×10^{-4}	0.0016	0.0039	0.0039	0.009	0.01	0.285							
σ/MPa	120	120	140	140	120	120	160	160	120	120	120	120			
t/ks	0	36	36	72	72	108	108	144	144	214.9	246.8	2430.8			
ε_c/%	0.053	0.068	0.09	0.109	0.09	0.104	0.159	0.181	0.124	0.148	0.151	0.179			
η	0.203	0.309	0.409	0.531	0.531	0.593	0.748	0.878	0.878	1.0	1.0	1.0			
ω	0	0.395	0.34	0.524	0.524	0.702	0.585	0.665	0.665	0.871	1.0	1.0			
q	1×10^{-6}	6.5×10^{-4}	0.0016	0.0039	0.0039	0.0049	0.017	0.024	0.026	0.028	0.03	0.285			

(continued)

Table 3.10 (continued)

σ/MPa	120	120	140	140	120	120	160	160	120	120	180	180	120	120	120
t/ks	0	36	36	72	72	108	108	144	144	180	180	216	216	243.3	2099.3
ε_c/%	0.053	0.068	0.09	0.109	0.09	0.104	0.159	0.181	0.124	0.137	0.237	0.25	0.149	0.151	0.167
η	0.203	0.309	0.409	0.531	0.531	0.593	0.748	0.878	0.878	0.878	1.0	1.0	1.0	1.0	1.0
ω	0	0.395	0.34	0.524	0.524	0.702	0.585	0.665	0.665	0.832	0.74	0.888	0.888	1.0	1.0
q	1×10^{-6}	6.5×10^{-4}	0.0016	0.0039	0.0039	0.0049	0.017	0.024	0.026	0.027	0.087	0.109	0.132	0.133	0.285

3.3.5 Stress-Rupture of SiC/SiC Composite with Stochastic Loading for Different Temperatures

Figures 3.12 and 3.13, and Tables 3.11 and 3.12 show the composite strain, interface debonding and oxidation ratio, and the broken fibers fraction of SiC/SiC composite corresponding to the loading sequences of Cases 1, 2, 3, and 4 for T = 850 and 900 °C in air atmosphere. When the temperature increases, the stress-rupture lifetime decreases and the time for the interface complete debonding and oxidation decreases. The increasing rate of the interface oxidation length and the degradation rate of the fiber strength increase with temperature, leading to the decrease of the stress-rupture lifetime and the time for the interface complete debonding and oxidation.

When the temperature is T = 850°C, the damage evolution and lifetime of SiC/SiC composite corresponding to loading spectrums of Cases 1, 2, 3, and 4 are analyzed, as follows:

- Under constant stress σ = 120 MPa, the composite's stress-rupture lifetime is t = 1705 ks; the time for the composite's interface complete debonding is t = 168.3 ks; the time for the composite's interface complete oxidation is t = 204.9 ks; the composite's failure strain is ε_c = 0.2%; and the corresponding broken fibers fraction is q = 0.285.
- For the stochastic loading Case 2, the stochastic loading stress of σ_s = 140 MPa occurs at t = 36 ks; the time interval is Δt = 36 ks; the composite's stress-rupture lifetime is t = 1703.9 ks; the time for the composite's interface complete debonding is t = 142.6 ks under the constant stress of σ = 120 MPa; the time for the interface complete oxidation is t = 179.9 ks under constant stress of σ = 120 MPa; the composite's failure strain is ε_c = 0.2%; and the corresponding broken fibers fraction is q = 0.285.
- For the stochastic loading Case 3, the stochastic loading stresses of σ_s = 140, 160 MPa occur at t = 36, 108 ks; the time interval is Δt = 36 ks; the composite's stress-rupture lifetime is t = 1679.4 ks; the time for the composite's interface complete debonding is t = 110.9 ks under stochastic loading stress of σ_s = 160 MPa; the time for the composite's interface complete oxidation is t = 171.2 ks under constant stress of σ = 120 MPa; the composite's failure strain is ε_c = 0.197%; and the corresponding broken fibers fraction is q = 0.285.
- For the stochastic loading Case 4, the stochastic loading stresses of σ_s = 140, 160, 180 MPa occur at t = 36, 108, 180 ks; the time interval is Δt = 36 ks; the composite's stress-rupture lifetime is t = 713.3 ks; the time for the composite's interface complete debonding is t = 110.9 ks under stochastic loading stress of σ_s = 160 MPa; the time for the composite's interface complete oxidation is t = 171.2 ks under constant stress σ = 120 MPa; the composite's failure strain is ε_c = 0.175%; and the corresponding broken fibers fraction is q = 0.285.

When the temperature is T = 900 °C, the damage evolution and lifetime of SiC/SiC composite corresponding to loading spectrums of Cases 1, 2, 3, and 4 are analyzed, as follows:

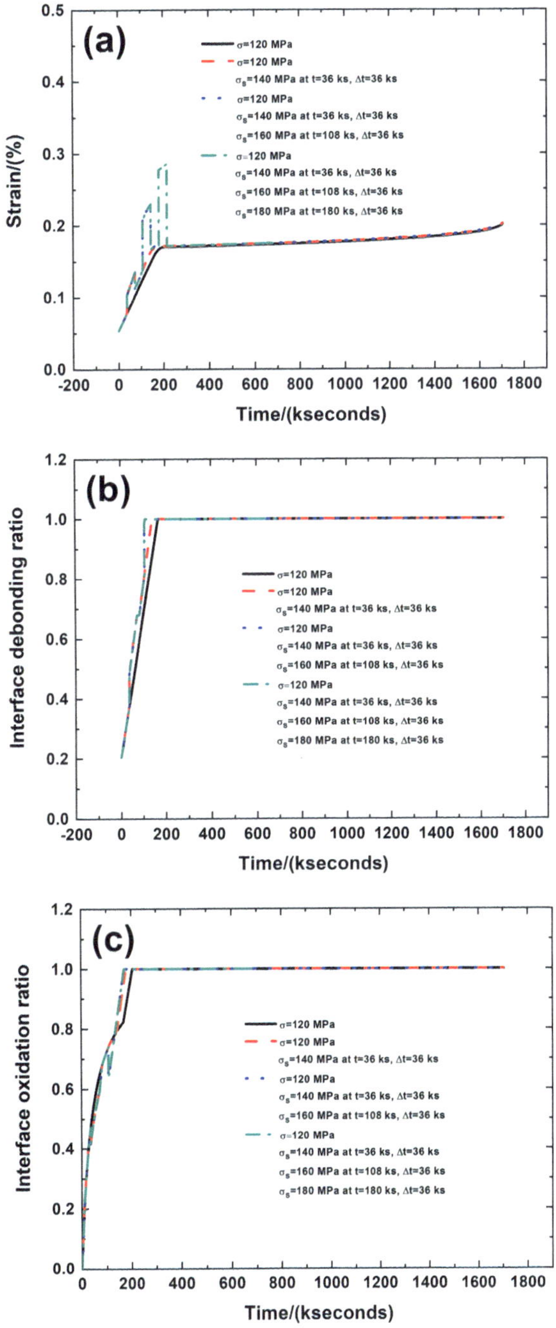

Fig. 3.12 **a** Composite strain versus the time curves; **b** the interface debonding ratio versus the time curves; **c** the interface oxidation ratio versus the time curves; and **d** the broken fibers fraction versus the time curves of SiC/SiC composite under stress-rupture loading of constant stress $\sigma = 120$ MPa, $\sigma_s = 140$, 140/160, 140/160/180 MPa at $t = 36$, 36/108, 36/108/180 ks and $\Delta t = 36$ ks when $T = 850$ °C

Fig. 3.12 (continued)

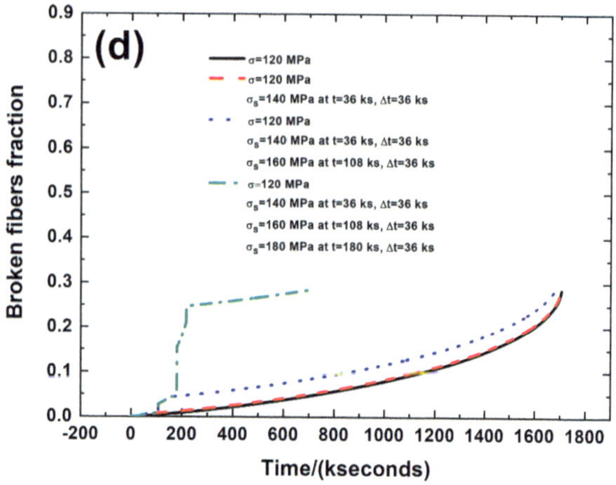

- Under constant stress σ = 120 MPa, the composite's stress-rupture lifetime is t = 1224.7 ks; the time for the composite's interface complete debonding is t = 120.5 ks; the time for the composite's interface complete oxidation is t = 146.8 ks; the composite's failure strain is $\varepsilon_c = 0.201\%$; and the corresponding broken fibers fraction is $q = 0.285$.
- For the stochastic loading Case 2, the stochastic loading stress of σ_s = 140 MPa occurs at t = 36 ks; the time interval is Δt = 36 ks; the composite's stress-rupture lifetime is t = 1223.4 ks; the time for the composite's interface complete debonding is t = 102.1 ks under the constant stress of σ = 120 MPa; the time for the composite's interface complete oxidation is t = 128.9 ks under constant stress of σ = 120 MPa; the composite's failure strain is $\varepsilon_c = 0.201\%$; and the corresponding broken fibers fraction is $q = 0.285$.
- For the stochastic loading Case 3, the stochastic loading stresses of σ_s = 140, 160 MPa occur at t = 36, 108 ks; the time interval is Δt = 36 ks; the composite's stress-rupture lifetime is t = 1183.6 ks; the time for the composite's interface complete debonding is t = 102.1 ks under constant stress of σ = 120 MPa; the time for the composite's interface complete oxidation is t = 122.6 ks under stochastic loading stress of σ_s = 160 MPa; the composite's failure strain is ε_c = 0.194%; and the corresponding broken fibers fraction is $q = 0.285$.
- For the stochastic loading Case 4, the stochastic loading stresses of σ_s = 140, 160, 180 MPa occur at t = 36, 108, 180 ks; the time interval is Δt = 36 ks; the composite's stress-rupture lifetime is t = 180 ks; the time for the composite's interface complete debonding is t = 102.1 ks under constant stress of σ = 120 MPa; and the time for the composite's interface complete oxidation is t = 122.6 ks under stochastic loading stress of σ_s = 160 MPa.

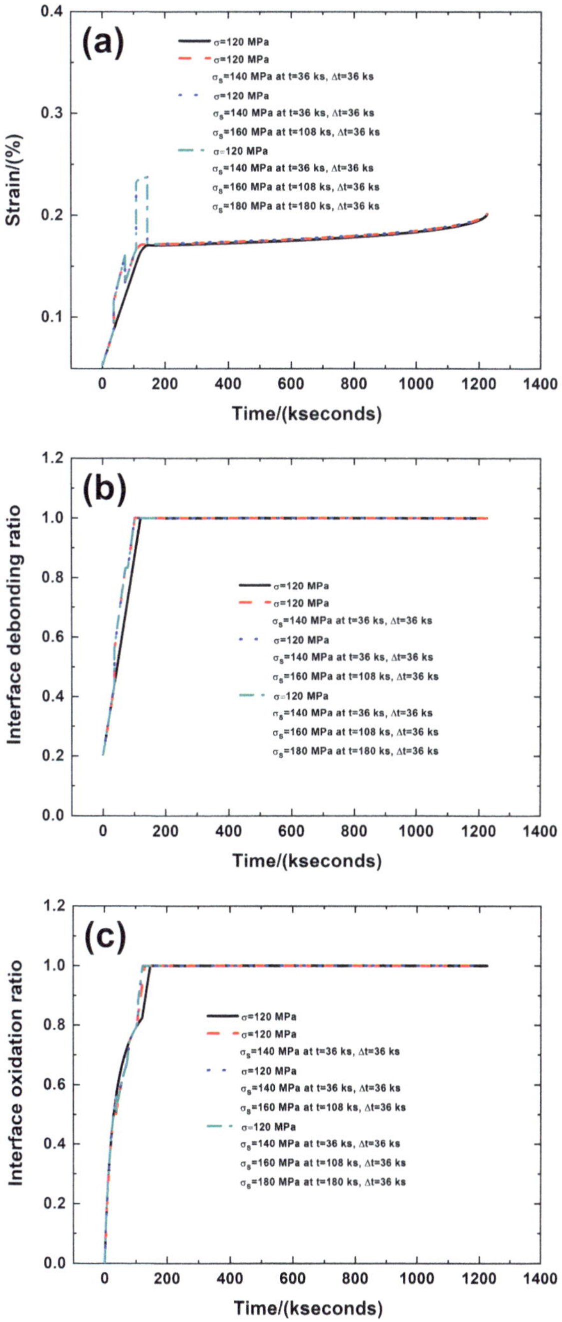

Fig. 3.13 **a** Composite strain versus the time curves; **b** the interface debonding ratio versus the time curves; **c** the interface oxidation ratio versus the time curves; and **d** the broken fibers fraction versus the time curves of SiC/SiC composite under stress-rupture loading of constant stress σ = 120 MPa, σ_s = 140, 140/160, 140/160/180 MPa at t = 36, 36/108, 36/108/180 ks and Δt = 36 ks when T = 900 °C

Fig. 3.13 (continued)

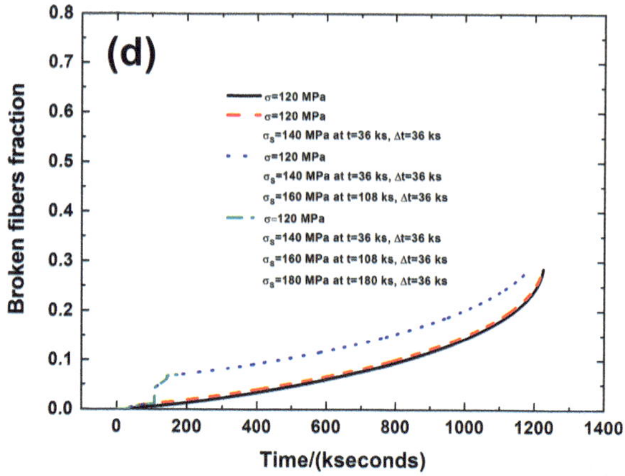

3.4 Experimental Comparisons

Lara-Curzio (1997) investigated the stress-rupture of SiC/SiC composite at T = 950 °C in air atmosphere. Experimental and theoretical predicted stress-rupture curves, interface debonding and interface oxidation, and the broken fibers fraction versus the time curves of SiC/SiC composite under constant and stochastic loading at T = 950 °C in air atmosphere are shown in Figs. 3.14 and 3.15 and Tables 3.13 and 3.14.

- Under the constant stress of σ = 80 MPa, the composite's strain increases from ε_c = 0.085 to 0.277% at t = 83.6 ks; the interface debonding ratio increases from η = 0.064 to 0.557 at t = 83.6 ks; the interface oxidation ratio increases from ω = 0 to 0.846 at t = 83.6 ks; and the broken fibers fraction increases from q = 0 to 0.291 at t = 83.6 ks. When the stochastic loading stress of σ_s = 90 MPa occurs at t = 18 ks and the time interval is Δt = 3.6 ks, the composite's stress-rupture lifetime is t = 82.6 ks; the composite's interface debonding ratio is η = 0.7; the composite's interface oxidation ratio is ω = 0.854; the composite's failure strain is ε_c = 0.327%; and the composite's broken fibers fraction is q = 0.284. When the stochastic loading stress of σ_s = 90 MPa occurs at t = 18, 25.2 ks and the time interval is Δt = 3.6 ks, the composite's stress-rupture lifetime is t = 75 ks; the composite's interface debonding ratio is η = 0.659; the composite's interface oxidation ratio is ω = 0.858; the composite's failure strain is ε_c = 0.293%; and the composite's broken fibers fraction is q = 0.285. When the stochastic loading stress of σ_s = 90 MPa occurs at t = 18, 25.2, 32.4 ks and the time interval is Δt = 3.6 ks, the composite's stress-rupture lifetime is t = 36 ks; the composite's interface debonding ratio is η = 0.371; the composite's interface oxidation ratio is ω = 0.711; the composite's failure strain is ε_c = 0.214%; and the corresponding broken fibers fraction is q = 0.229.
- Under the constant stress σ = 100 MPa, the composite's strain increases from ε_c = 0.1 to 0.2% at t = 33.7 ks; the composite's interface debonding ratio increases

Table 3.11 Composite strain, fiber/matrix interface debonding and oxidation ratio, and broken fibers fraction of SiC/SiC composite under stress-rupture loading of constant stress $\sigma = 120$ MPa, $\sigma_s = 140$, 140/160, 140/160/180 MPa at $t = 36$, 36/108, 36/108/180 ks and $\Delta t = 36$ ks at $T = 850$ °C in air atmosphere

σ/MPa	120	120	120	120	120									
t/ks	0	36	168.3	204.9	1705									
ε_c/%	0.053	0.077	0.163	0.17	0.2									
η	0.203	0.374	1.0	1.0	1.0									
ω	0	0.472	0.822	1.0	1.0									
q	1×10^{-6}	0.001	0.007	0.009	0.285									
σ/MPa	120	120	140	140	120	120	120	120						
t/ks	0	36	36	72	72	142.6	179.9	1703.9						
ε_c/%	0.053	0.077	0.103	0.135	0.112	0.162	0.171	0.2						
η	0.203	0.374	0.483	0.678	0.678	1.0	1.0	1.0						
ω	0	0.472	0.415	0.592	0.592	0.793	1.0	1.0						
q	1×10^{-6}	0.001	0.002	0.006	0.006	0.009	0.011	0.285						
σ/MPa	120	120	140	140	120	120	160	160	120	120	120			
t/ks	0	36	36	72	72	108	108	144	144	171.2	1679.4			
ε_c/%	0.053	0.077	0.103	0.135	0.112	0.138	0.206	0.229	0.166	0.171	0.197			
η	0.203	0.374	0.483	0.678	0.678	0.814	0.983	1.0	1.0	1.0	1.0			
ω	0	0.472	0.415	0.592	0.592	0.739	0.643	0.842	0.842	1.0	1.0			
q	1×10^{-6}	0.001	0.002	0.006	0.006	0.0078	0.027	0.039	0.042	0.044	0.285			

(continued)

Table 3.11 (continued)

σ/MPa	120	120	140	140	120	120	160	160	120	120	180	180	120	120
t/ks	0	36	36	72	72	108	108	144	144	180	180	216	216	713.3
ε_c/%	0.053	0.077	0.103	0.135	0.112	0.138	0.206	0.229	0.166	0.171	0.277	0.286	0.172	0.175
η	0.203	0.374	0.483	0.678	0.678	0.814	0.983	1.0	1.0	1.0	1.0	1.0	1.0	1.0
ω	0	0.472	0.415	0.592	0.592	0.739	0.643	0.842	0.842	1.0	1.0	1.0	1.0	1.0
q	1×10^{-6}	0.001	0.002	0.006	0.006	0.0078	0.027	0.039	0.042	0.044	0.154	0.21	0.247	0.285

Table 3.12 Composite strain, interface debonding and oxidation ratio, and broken fibers fraction of SiC/SiC composite under stress-rupture loading of constant stress $\sigma = 120$ MPa, $\sigma_s = 140$, 140/160, 140/160/180 MPa at $t = 36$, 36/108, 36/108/180 ks and $\Delta t = 36$ ks at $T = 900$ °C in air atmosphere

σ/MPa	120	120	120	120	120						
t/ks	0	36	120.5	146.8	1224.7						
ε_c/%	0.053	0.087	0.163	0.17	0.201						
η	0.203	0.442	1.0	1.0	1.0						
ω	0	0.558	0.822	1.0	1.0						
q	1×10^{-6}	0.0015	0.007	0.009	0.285						
σ/MPa	120	120	140	140	120	120	120	120			
t/ks	0	36	36	72	72	102.1	128.9	1223.4			
ε_c/%	0.053	0.087	0.116	0.161	0.134	0.162	0.171	0.201			
η	0.203	0.442	0.561	0.833	0.833	1.0	1.0	1.0			
ω	0	0.558	0.5	0.673	0.673	0.794	1.0	1.0			
q	1×10^{-6}	0.0015	0.0039	0.0094	0.0094	0.011	0.013	0.285			
σ/MPa	120	120	140	140	120	120	160	160	120	120	
t/ks	0	36	36	72	72	108	108	144	144	1183.6	
ε_c/%	0.053	0.087	0.116	0.161	0.134	0.166	0.232	0.237	0.171	0.194	
η	0.203	0.442	0.561	0.833	0.833	1.0	1.0	1.0	1.0	1.0	
ω	0	0.558	0.5	0.673	0.673	0.839	0.882	1.0	1.0	1.0	
q	1×10^{-6}	0.0015	0.0039	0.0094	0.0094	0.012	0.043	0.062	0.067	0.285	
σ/MPa	120	120	140	140	120	120	160	160	120	120	180
t/ks	0	36	36	72	72	108	108	144	144	180	180
ε_c/%	0.053	0.087	0.116	0.161	0.134	0.166	0.232	0.237	0.171	0.172	–
η	0.203	0.442	0.561	0.833	0.833	1.0	1.0	1.0	1.0	1.0	1.0
ω	0	0.558	0.5	0.673	0.673	0.839	0.882	1.0	1.0	1.0	1.0
q	1×10^{-6}	0.0015	0.0039	0.0094	0.0094	0.012	0.043	0.062	0.067	0.07	–

from $\eta = 0.078$ to 0.28 at $t = 33.7$ks; the composite's interface oxidation ratio increases from $\omega = 0$ to 0.617 at $t = 33.7$ ks; and the corresponding broken fibers fraction increases from $q = 1 \times 10^{-6}$ to 0.228 at $t = 33.7$ ks. When the stochastic loading stress of $\sigma_s = 100$ MPa occurs at $t = 7.2$ ks and the time interval is $\Delta t = 3.6$ ks, the composite's stress-rupture lifetime is $t = 32.3$ ks; the composite's interface debonding ratio is $\eta = 0.343$; the composite's interface oxidation ratio is $\omega = 0.627$; the composite's failure strain is $\varepsilon_c = 0.224\%$; and the corresponding broken fibers fraction is $q = 0.285$. When the stochastic loading stress of $\sigma_s = 100$ MPa occurs at $t = 7.2$ ks and the time interval is $\Delta t = 5.4$ ks, the composite's

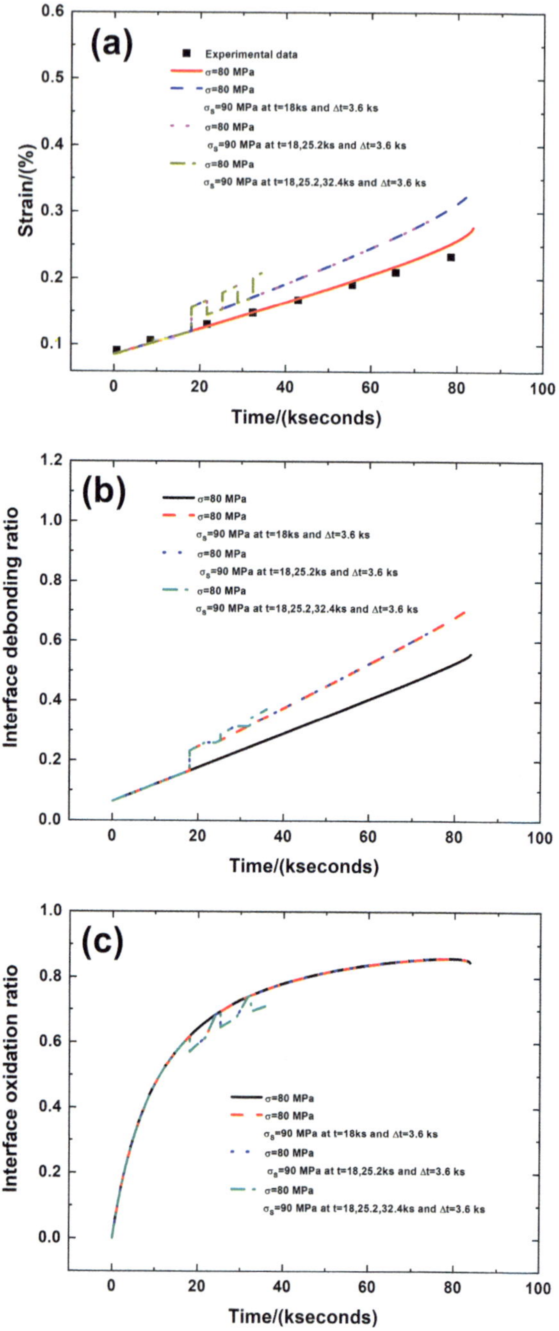

Fig. 3.14 **a** Composite's strain versus the time curves; **b** the composite's interface debonding ratio versus the time curves; **c** the composite's interface oxidation ratio versus the time curves; and **d** the broken fibers fraction versus the time curves of SiC/SiC composite under $\sigma =$ 80 MPa, $\sigma_s = 90$ MPa at $t =$ 18, 18/25.2, 18/25.2/32.4 ks and $\Delta t = 3.6$ ks at $T =$ 950 °C in air atmosphere

Fig. 3.14 (continued)

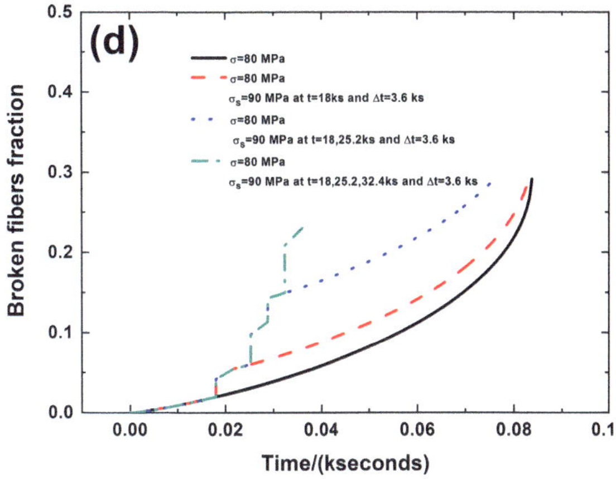

stress-rupture lifetime is $t = 29.8$ ks; the composite's interface debonding ratio is $\eta = 0.318$; the composite's interface oxidation ratio is $\omega = 0.623$; the composite's failure strain is $\varepsilon_c = 0.212\%$; and the corresponding broken fibers fraction is $q = 0.285$.

3.5 Conclusions

In this chapter, the damage evolution and lifetime of fiber-reinforced CMCs under stress-rupture with stochastic loading at intermediate temperatures were investigated. Damage evolution of the composite's strain, interface debonding and interface oxidation ratio, and the broken fibers fraction versus the time curves of SiC/SiC composite subjected to constant stress and three different stochastic loading conditions were analyzed. For the constant stress and stochastic loading Cases 2, 3, and 4:

- When the fiber volume increased, the stress-rupture lifetime increased, the time for the interface complete debonding increased, and the failure strain decreased.
- When the saturation matrix crack spacing increased, the time for the interface complete debonding and oxidation increased and the failure strain decreased.
- When the fiber/matrix interface debonding energy increased, the time for the interface complete debonding increased and the failure strain decreased.
- When the fiber/matrix interface shear stress in the slip region increased, the stress-rupture lifetime increased and the time for the interface complete debonding increased; when the fiber/matrix interface shear stress in the oxidation region increased, the time for the interface complete debonding increased and the failure strain decreased.
- When the temperature increased, the stress-rupture lifetime decreased and the time for the interface complete debonding and oxidation decreased.

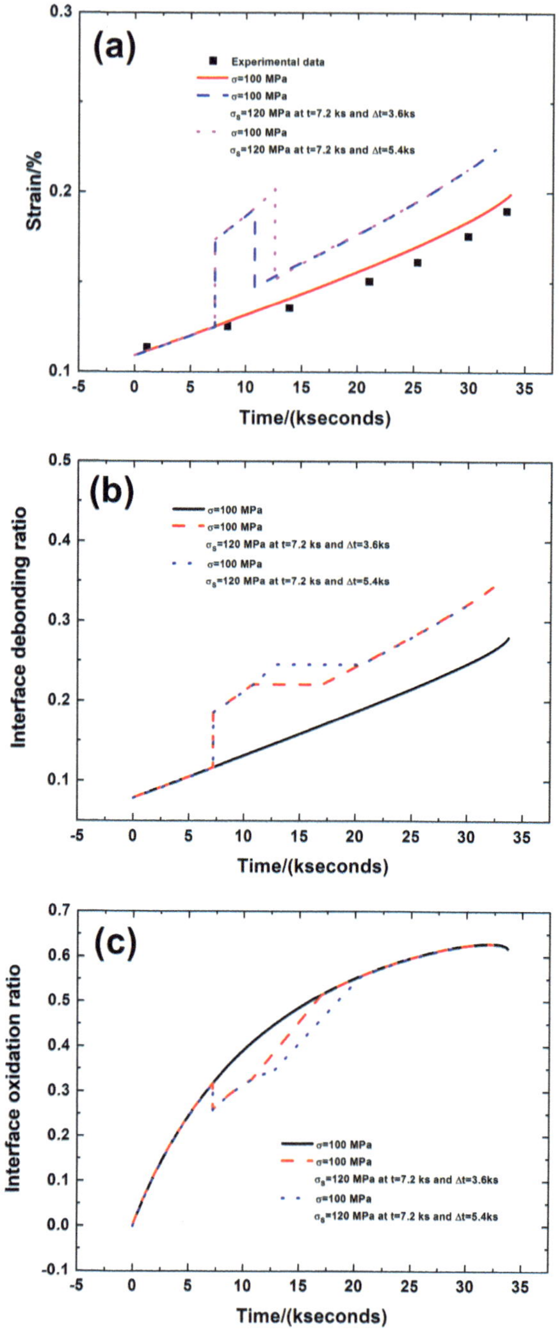

Fig. 3.15 **a** Composite's strain versus the time curves; **b** the composite's interface debonding ratio versus the time curves; **c** the composite's interface oxidation ratio versus the time curves; and **d** the broken fibers fraction versus the time curves of SiC/SiC composite under $\sigma =$ 100 MPa, $\sigma_s = 120$ MPa at $t = 7.2$ ks and $\Delta t = 3.6$ and 5.4 ks at $T = 950$ °C in air atmosphere

Fig. 3.15 (continued)

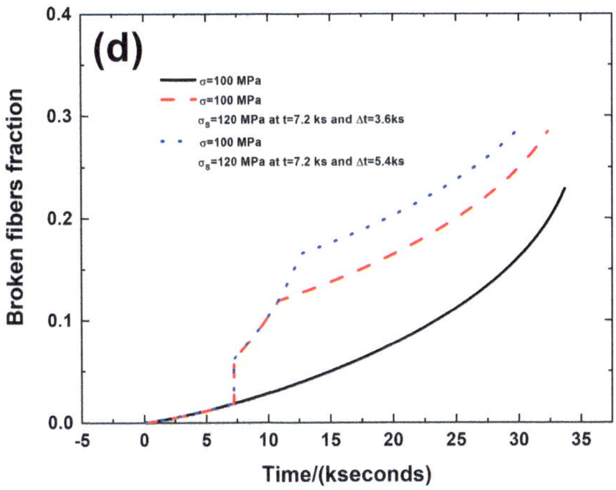

Table 3.13 Composite's strain, interface debonding and oxidation ratio, and broken fibers fraction of SiC/SiC composite under $\sigma = 80$ MPa, $\sigma_s = 90$ MPa at $t = 18$, 18/25.2, 18/25.2/32.4 ks and $\Delta t = 3.6$ ks at $T = 950$ °C in air atmosphere

σ/MPa	80	80	80									
t/ks	0	18	83.6									
ε_c/%	0.085	0.12	0.277									
η	0.064	0.166	0.557									
ω	0	0.617	0.846									
q	0	0.019	0.291									
σ/MPa	80	80	90	90	80	120						
t/ks	0	18	18	21.6	21.6	82.6						
ε_c/%	0.085	0.12	0.156	0.167	0.145	0.327						
η	0.064	0.166	0.232	0.259	0.259	0.7						
ω	0	0.617	0.57	0.612	0.612	0.854						
q	0	0.019	0.04	0.054	0.054	0.284						
σ/MPa	80	80	90	90	80	80	90	90	80	80		
t/ks	0	18	18	21.6	21.6	25.2	25.2	28.8	28.8	75		
ε_c/%	0.085	0.12	0.156	0.167	0.145	0.153	0.178	0.189	0.163	0.293		
η	0.064	0.166	0.232	0.259	0.259	0.267	0.287	0.314	0.314	0.659		
ω	0	0.617	0.57	0.612	0.612	0.693	0.645	0.672	0.672	0.858		
q	0	0.019	0.04	0.054	0.054	0.059	0.097	0.113	0.142	0.285		
σ/MPa	80	80	90	90	80	80	90	90	80	80	90	90
t/ks	0	18	18	21.6	21.6	25.2	25.2	28.8	28.8	32.4	32.4	36
ε_c/%	0.085	0.12	0.156	0.167	0.145	0.153	0.178	0.189	0.163	0.171	0.201	0.214
η	0.064	0.166	0.232	0.259	0.259	0.267	0.287	0.314	0.314	0.319	0.342	0.371
ω	0	0.617	0.57	0.612	0.612	0.693	0.645	0.672	0.672	0.743	0.694	0.711
q	0	0.019	0.04	0.054	0.054	0.059	0.097	0.113	0.142	0.149	0.208	0.229

Table 3.14 Composite's strain, interface debonding and oxidation ratio, and broken fibers fraction of SiC/SiC composite under σ = 100 MPa, σ_s = 120 MPa at t = 7.2 ks and Δt = 3.6, 5.4 ks at T = 950 °C in air atmosphere

σ/MPa	100	100	100			
t/ks	0	7.2	33.7			
ε_c/%	0.109	0.125	0.198			
η	0.078	0.116	0.279			
ω	0	0.317	0.617			
q	1×10^{-6}	0.018	0.228			
σ/MPa	100	100	120	120	100	100
t/ks	0	7.2	7.2	10.8	10.8	32.3
ε_c/%	0.109	0.125	0.174	0.191	0.147	0.224
η	0.078	0.116	0.186	0.22	0.22	0.343
ω	0	0.317	0.258	0.326	0.326	0.627
q	1×10^{-6}	0.018	0.06	0.119	0.119	0.285
σ/MPa	100	100	120	120	100	100
t/ks	0	7.2	7.2	12.6	12.6	29.8
ε_c/%	0.109	0.125	0.174	0.201	0.151	0.212
η	0.078	0.116	0.186	0.246	0.246	0.318
ω	0	0.317	0.258	0.342	0.342	0.623
q	1×10^{-6}	0.018	0.06	0.164	0.164	0.285

References

Casas L, Martinez-Esnaola JM (2003) Modeling the effect of oxidation on the creep behavior of fiber-reinforced ceramic matrix composites. Acta Mater 51:3745–3757. https://doi.org/10.1016/S1359-6454(03)00189-7

Curtin WA (1991) Theory of mechanical properties of ceramic-matrix composites. J Am Ceram Soc 74:2837–2845. https://doi.org/10.1111/j.1151-2916.1991.tb06852.x

Dassios KG, Aggelis DG, Kordatos EZ, Matikas TE (2013) Cyclic loading of a SiC-fiber reinforced ceramic matrix composite reveals damage mechanisms and thermal residual stress state. Compos Part A 44:105–113. https://doi.org/10.1016/j.compositesa.2012.06.011

Gao Y, Mai Y, Cotterell B (1988) Fracture of fiber-reinforced materials. J Appl Math Phys 39:550–572. https://doi.org/10.1007/BF00948962

Gauthier W, Lamon J (2009) Delayed failure of Hi-Nicalon and Hi-Nicalon S multifilament tows and single filaments at intermediate temperatures (500–800°C). J Am Ceram Soc 92:702–709. https://doi.org/10.1111/j.1551-2916.2009.02924.x

Godin N, Reynaud P, Fantozzi G (2019) Contribution of AE analysis in order to evaluate time to failure of ceramic matrix composites. Eng Fract Mech 210:452–469. https://doi.org/10.1016/j.engfracmech.2018.03.006

Halverson HG, Curtin WA (2002) Stress rupture in ceramic-matrix composites: theory and Experiment. J Am Ceram Soc 85:1350–1365. https://doi.org/10.1111/j.1151-2916.2002.tb00279.x

Ikarashi Y, Ogasawara T, Aoki T (2019) Effects of cyclic tensile loading on the rupture behavior of orthogonal 3-D woven SiC fiber/SiC matrix composites at elevated temperatures in air. J Euro Ceram Soc 39:806–812. https://doi.org/10.1016/j.jeurceramsoc.2018.10.016

Lara-Curzio E (1997) Stress rupture of Nicalon/SiC continuous fiber ceramic composites in air at 950°C. J Am Ceram Soc 80:3268–3272. https://doi.org/10.1111/j.1151-2916.1997.tb03266.x

Lara-Curzio E (1999) Analysis of oxidation-assisted stress-rupture of continuous fiber-reinforced ceramic matrix composites at intermediate temperatures. Compos Part A 30:549–554. https://doi.org/10.1016/S1359-835X(98)00148-1

Li LB (2017) Damage evolution of cross-ply ceramic-matrix composites under stress-rupture and cyclic loading at elevated temperatures in oxidizing atmosphere. Mater Sci Eng A 688:315–321. https://doi.org/10.1016/j.msea.2017.02.012

Li LB (2020) Damage evolution and lifetime of fiber-reinforced ceramic-matrix composites under stress-rupture with stochastic loading at intermediate temperatures. Mater Today Commun 23:100940. https://doi.org/10.1016/j.mtcomm.2020.100940

Maillet E, Godin N, R'Mili M, Reynaud P, Lamon J, Fantozzi G (2012) Analysis of acoustic emission energy release during static fatigue tests at intermediate temperatures on ceramic matrix composites: towards rupture time prediction. Comp Sci Technol 72:1001–1007. https://doi.org/10.1016/j.compscitech.2012.03.011

Morscher GN (1997) Tensile stress rupture of SiC_f/SiC_m minicomposites with carbon and boron nitride interphases at elevated temperatures in air. J Am Ceram Soc 80:2029–2042. https://doi.org/10.1111/j.1151-2916.1997.tb03087.x

Rebillat F (2014) Advances in self-healing ceramic matrix composites. In: Low IM (ed) Advances in ceramic matrix composites, Woodhead Publishing

Recle E, Godin N, Reynaud P, Fantozzi G (2017) Fatigue lifetime of ceramic matrix composites at intermediate temperature by acoustic emission. Materials 10:658. https://doi.org/10.3390/ma10060658

Ruggles-Wrenn MB, Christensen DT, Chamberlain AL, Lane JE, Cook TS (2011) Effect of frequency and environment on fatigue behavior of a CVI SiC/SiC ceramic matrix composite at 1200°C. Compos Sci Technol 71:190–196. https://doi.org/10.1016/j.compscitech.2010.11.008

Verrilli MJ, Opila EJ, Calomino A, Kiser JD (2004) Effect of environment on the stress-rupture behavior of a carbon-fiber-reinforced silicon carbide ceramic matrix composites. J Am Ceram Soc 87:1536–1542. https://doi.org/10.1111/j.1551-2916.2004.01536.x

Chapter 4
Fatigue Life of Ceramic-Matrix Composites Subjected to Stochastic Loading at Elevated Temperature

Abstract In this chapter, a micromechanical approach is developed to predict fatigue life of fiber-reinforced ceramic-matrix composites (CMCs) subjected to stochastic overloading stress at elevated temperature. Multiple elevated temperature fatigue damage mechanisms of interface wear and interface and fiber oxidation are considered in the analysis of elevated temperature fatigue damage evolution and fracture under stochastic overloading. Relationships between stochastic overloading stress level and occurrence applied cycle number, broken fibers fraction, and fatigue life decreasing rate are established. Experimental fatigue life S–N curves of C/SiC composites with different fiber preforms (i.e., unidirectional, cross-ply, 2D woven, 2.5D woven, and 3D braided) are predicted for different stochastic overloading stress levels and occurrence applied cycle number.

Keywords Ceramic-matrix composites (CMCs) · Stochastic load · Fatigue · Life prediction

4.1 Introduction

Ceramic-matrix composite (CMC) has the characteristics of low density, high hardness, excellent thermal stability, and strong chemical resistance (Naslain 2004). Its density is only 1/3 of that of superalloy, and its strength is twice as high as that of superalloy. It can withstand high temperature from 1000 to 1500 °C (200−240 °C higher than that of superalloy), and its structural durability is better. At the same time, the inherent fracture toughness and damage tolerance of CMC are high, which is suitable for the hot-section components of gas turbine engine, and can operate at higher turbine inlet temperature and less cooling air (more than 1300 °C), and the engine efficiency and fuel consumption rate are significantly improved (Ohnabe et al. 1999). Aeroengine manufacturers attach great importance to the development of CMC technology and strive to introduce this material into hot-section components such as combustion chamber liners, nozzle guide vanes,5 and even turbine rotor parts (Zhu 2018). In the early 1980s, SNECMA company of France carried out research on the application of CMCs to meet the requirements of high-temperature materials for the nozzle of aeroengine and developed CERASEPR A300 SiC/SiC composite.

L. Li, *Damage and Fracture of Ceramic-Matrix Composites Under Stochastic Loading*, Advanced Ceramics and Composites 2,
https://doi.org/10.1007/978-981-16-2141-3_4

In the 1990s, CERASEPR A410 SiC/SiC composite with self-healing technology was developed. At present, the nozzle inner adjusting plate test has been successfully completed on M88-2E4 developed engine. Subsequently, in the mid-1990s, SNECMA company and P&W company carried out the engineering research and application of SiC/SiC composite on the nozzle components of F100 engine. The seal adjusting plate of CERASEPR A410 nozzle was installed on F100-PW-229 and F100-PW-220 aeroengines, and the SiC/SiC mixer was applied to CFM56 commercial aeroengine (Li 2020). The LEAP-1A civil turbofan engine equipped with CMC high-pressure turbine component has successfully completed its first flight on the new Airbus A320neo aircraft on May 19, 2015, which indicates that CMC has made a new breakthrough in the application of aeroengine hot-section components. At NASA N+3 advanced engine project, new generation CMC combustion chamber, high-pressure turbine guide vane and blade, low-pressure turbine blade, and high-pressure turbine support ring are developed (Li 2020). Since the application of CMCs involves components with lives that are measured in tens of thousands of hours, successful design and implementation of CMC components depend on the knowledge of material behavior over periods of time comparable to the expected service life of component (Li et al. 2020; Ruggles-Wrenn et al. 2011). To ensure the reliability and safety of CMC components, it is necessary to develop performance evaluation, damage evolution, strength, and life prediction tools for airworthiness certification (Li 2019a).

Under cyclic fatigue loading at elevated temperature, multiple fatigue damage mechanisms of matrix cracking, interface debonding, and wear and fibers fracture degrade the mechanical performance of fiber-reinforced CMCs (Zhu 2018). Quiney et al. (Quiney et al. 2020) investigated fatigue damage in a SiC/SiC composite using in situ X-ray computed tomography (XCT) at room temperature. Damage in the forms of matrix cracks has been visualized using in situ XCT inspection of fatigue test coupons. The closure of matrix cracks has also been observed under cyclic loading and unloading. Ruggles-Wrenn et al. (2012) investigated tension–tension fatigue behavior of a Hi-Nicalon™ SiC/[SiC-B_4C] composite at T = 1200 °C in air and in steam atmospheres. Presence of steam had little influence on fatigue performance at loading frequency f = 1.0 Hz, however, noticeably degraded the fatigue lifetimes at f = 0.1 Hz. Prior fatigue in air and in steam caused significant reduction in composite tensile strength and modulus. Li (2015a,b) investigated tension–tension fatigue behavior of unidirectional and cross-ply C/SiC composites at T = 800 °C in air atmosphere. Compared with unidirectional C/SiC composite, the fatigue damage and fracture processes were much more complicated for cross-ply C/SiC composite due to transverse cracking in 90° plies. Hysteresis loops modulus and area were used to characterize fatigue damage evolution and fracture of unidirectional and cross-ply C/SiC composites. Microstructure observations of matrix cracks and fibers fracture and pullout were conducted to relate the microdamage mechanisms with macrofatigue damage degradation and evolution. Mall and Engesser (2006) investigated tension–tension fatigue behavior of 2D C/SiC composites at T = 550 °C in air atmosphere corresponding to different fatigue loading frequencies. Oxidation of carbon fibers was the major reason for the fatigue failure of C/SiC composites at elevated temperature and caused a reduction in fatigue life of C/SiC composite at $T = 550$ °C under low loading frequency compared with that at room temperature. Yang (2011) investigated tension–tension fatigue behavior of 2.5D C/SiC composite at $T = 800$ °C in air atmosphere. Fatigue damage development of C/SiC composite at $T = 800$ °C was compared with that at room temperature. The fatigue life decreased rapidly compared with that at room temperature. Du et al. (2002), Du and Qiao (2011) investigated tension–tension fatigue behavior of 3D C/SiC composite at room temperature, 1300 and 1500 °C in inert atmosphere. At room temperature, the fatigue limit stress was approximately $\sigma_{\text{limit}} = 235$ MPa, which was 85% of composite tensile strength; at $T = 1300$ °C in vacuum atmosphere, the composite fatigue limit stress was approximately $\sigma_{\text{limit}} = 285$ MPa, which was approximately 94% of composite tensile strength; at $T = 1500$ °C in vacuum atmosphere, the composite fatigue limit stress was approximately $\sigma_{\text{limit}} = 240$ MPa, which was approximately 92% of composite tensile strength. After fatigue fracture at $T = 1300$ °C, the length of fibers pullout is much less than that at room temperature. After fatigue fracture at $T = 1500$ °C, fibers pullout length is long for the condition under low stress ratio, high loading frequency, and high cycle number. Liu et al. (2020) developed a continuum damage constitutive model to predict damage evolution of oxide/oxide CMCs under multi-axial cyclic loads. The constitutive model can predict the stress–strain response and stiffness degradation of oxide/oxide CMCs under monotonic and cyclic load. Li (2016a, b, 2015) developed a micromechanical approach to predict fatigue life of CMCs at elevated temperature. The degradation rate of fiber/matrix interface shear stress and fibers strength affects the fatigue life. However, stochastic overloading occurs during applications, which affects the fatigue damage and lifetime of CMCs (Suzuki et al. 2019). Li (2019) developed a micromechanical approach to predict stress-rupture lifetime of 2D woven SiC/SiC composite at intermediate temperature. However, in the researches mentioned above, effect of stochastic overloading stress on fatigue life of fiber-reinforced CMCs at elevated temperature has not been investigated.

In this chapter, a micromechanical approach to predict fatigue life of fiber-reinforced CMCs subjected to stochastic overloading is developed. Multiple elevated temperature fatigue damage mechanisms of interface wear and interface and fibers oxidation are considered in the analysis of fatigue damage evolution and fracture of CMCs under stochastic overloading. Relationships between stochastic overloading stress and occurrence applied cycle number, broken fibers fraction, and fatigue life decreasing rate are established. Experimental fatigue life S–N curves of C/SiC composites with different fiber preforms are predicted for different stochastic overloading stress levels and occurrence cycle number.

4.2 Theoretical Model

When stochastic overloading stress occurs under cyclic fatigue loading, multiple fatigue damage mechanisms of matrix cracking, interface debonding, and fibers failure occur inside of CMCs (Min et al. 2014; Reynaud 1996; Mizuno et al. 1996).

Figure 4.1 shows the schematic of stochastic overloading stress under cyclic fatigue loading. Stochastic overloading stress σ_s occurs at different applied cycle numbers N_1, N_2, and N_3.

At elevated temperature, oxidation decreases fibers strength and increases the interface oxidation length. The cyclic-dependent degradation rate of the fiber/matrix interface shear stress Θ and fibers strength Ω can be determined by Eqs. (4.1) and (4.2) (Evans et al. 1995; Lee and Stinchcomb 1994)

$$\Theta(N) = \frac{1}{1 - (1 - \varphi)\left[1 - \exp\left(-\omega N^{\lambda}\right)\right]} \tag{4.1}$$

$$\Omega(N) = \frac{1}{1 - p_1(\log N)^{p_2}} \tag{4.2}$$

where N is the cycle number, φ is the ratio between steady interface shear stress and initial interface shear stress, ω and λ are the interface wear model parameter, and p_1 and p_2 are the fiber strength degradation model parameter.

The time-dependent fibers strength degradation can be determined by Eq. (4.3) (Lara-Curzio 1999)

$$\sigma_0(t) = \begin{cases} \sigma_0, t \le \frac{1}{k}\left(\frac{K_{IC}}{Y\sigma_0}\right)^4 \\ \frac{K_{IC}}{Y\sqrt[4]{kt}}, t > \frac{1}{k}\left(\frac{K_{IC}}{Y\sigma_0}\right)^4 \end{cases} \tag{4.3}$$

where K_{IC} is the critical stress intensity factor, Y is a geometric parameter, k is the parabolic rate constant, and μ_1 and μ_2 are fitting parameters dependent on temperature.

When fibers failure occurs, stress distributions between the intact and broken fibers can be determined using the global load sharing (GLS) criterion. Relationships between stochastic overloading stress σ_s, intact fiber stress Φ_s, and broken fibers stress Φ_b can be determined by Eq. (4.4) (Curtin 1991).

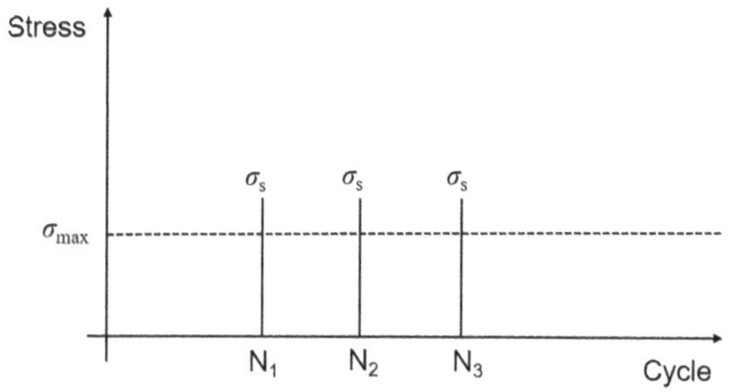

Fig. 4.1 Schematic of stochastic overloading under cyclic fatigue loading

$$\frac{\sigma_s}{V_f} = \left[1 - P_f\left(1 + \frac{2l_f}{l_c}\right)\right]\Phi_s + P_r\frac{2l_f}{l_c}\Phi_b \tag{4.4}$$

where

$$P_f = \alpha[\beta P_{fa} + (1 - \beta)P_{fb}] + P_{fc} + P_{fd} \tag{4.5}$$

$$P_r = P_{fc} + P_{fd} \tag{4.6}$$

where V_f is the volume fraction of fibers in the composite, l_c is the matrix crack spacing, l_f is the fiber sliding length, α is the oxidation fraction among multiple matrix cracks, β is the oxidation fiber fraction in the oxidized region, and P_{fa}, P_{fb}, P_{fc}, and P_{fd} are the fiber failure probability of oxidized fiber in the oxidation region, unoxidized fiber in the oxidation region, fiber in the interface debonding region, and interface bonding region.

$$P_{fa}(\sigma_s) = 1 - \exp\left\{-2\frac{l_t}{l_0}\left[\frac{\Phi_s}{\sigma_0(t)}\right]^m\right\} \tag{4.7}$$

$$P_{fb}(\sigma_s) = 1 - \exp\left\{-2\frac{l_t}{l_0}\left(\frac{\Phi_s}{\sigma_0}\right)^m\right\} \tag{4.8}$$

$$P_{fc}(\sigma_s) = 1 - \exp\left\{-\frac{r_f(\Phi_s)^{m+1}}{l_0\left(\frac{\sigma_0}{\Omega}\right)^m\left(\frac{\tau_{io}}{\Theta}\right)(m+1)}\left[1 - \left(1 - \frac{l_d}{l_f}\right)^{m+1}\right]\right\} \tag{4.9}$$

$$P_{fd}(\sigma_s) = 1 - \exp\left\{-\frac{2r_f(\Phi_s)^m}{\rho l_0\left(\frac{\sigma_0}{\Omega}\right)^m(m+1)\left(1 - \frac{\sigma_{fo}}{\Phi_s} - \frac{l_d}{l_s}\right)}\left[1 - \frac{l_d}{l_f} - \left(1 - \frac{\sigma_{fo}}{\Phi_s} - \frac{l_d}{l_f}\right)\frac{\rho l_d}{r_f}\right)^{m+1}\right.$$
$$\left.\left. - \left(1 - \frac{l_d}{l_f} - \left(1 - \frac{\sigma_{fo}}{\Phi_s} - \frac{l_d}{l_f}\right)\frac{\rho l_c}{2r_f}\right)^{m+1}\right]\right\} \tag{4.10}$$

where r_f is the fiber radius, m is the fiber Weibull modulus, l_d is the interface debonding length, and ρ is the shear-lag model parameter (Budiansky et al. 1986).

Using Eqs. (4.4)–(4.6), the intact fiber stress under stochastic overloading stress can be obtained. Substituting the intact fibers stress into Eqs. (4.7)–(4.10), the fibers failure probability in different damage regions can be obtained. When the fibers failure probability approaches critical value, the composite fatigue fracture occurs.

The fatigue life decreasing rate is defined by Eq. (4.11).

$$\Lambda = \frac{N_f(\sigma_{max}) - N_f(\sigma_s)}{N_f(\sigma_{max})} \tag{4.11}$$

where $N_f(\sigma_{max})$ is the fatigue failure cycle number under constant fatigue peak stress, and $N_f(\sigma_s)$ is the fatigue failure cycle number under stochastic overloading stress.

4.3 Experimental Comparisons

In this section, using the developed micromechanical fatigue life prediction approach, the elevated temperature fatigue life S–N curves of unidirectional, cross-ply, 2D woven, 2.5D woven, and 3D braided C/SiC composites under different stochastic overloading stress levels are predicted. Material properties and model parameters of C/SiC composite are listed in Table 4.1. Effects of stochastic overloading stress and occurrence applied cycle number on fatigue life, broken fibers fraction, and fatigue life decreasing rate of different C/SiC composites are analyzed.

4.3.1 Fatigue Life Prediction of Unidirectional C/SiC Composite Under Different Stochastic Overloading Stresses at T = 800 °C in Air Atmosphere

Li (2015a) performed experimental investigations on tension–tension fatigue behavior of unidirectional C/SiC composite at $T = 800$ °C in air atmosphere. The unidirectional $T - 700^{TM}$ carbon fiber-reinforced silicon carbide composite was fabricated using the hot-pressing (HT) method. The fibers volume fraction in the composite was approximately $V_f = 0.4$, and the monotonic tensile composite strength was approximately $\sigma_{uts} = 320$ MPa at $T = 800$ °C. The fatigue loading stress ratio was $R = 0.1$, and the fatigue loading frequency was $f = 10$ Hz. The fatigue limit stress was approximately $\sigma_{limit} = 22\%\sigma_{uts}$.

Figure 4.2a shows experimental and predicted fatigue life S–N curves of unidirectional C/SiC composite at $T = 800$ °C in air atmosphere under different stochastic

Table 4.1 Material properties of C/SiC composite

Items	Unidirectional	Cross-ply	2D woven	2.5D woven	3D braided	3D braided
Manufacturing process	Hot pressing	Hot Pressing	CVI	CVI	CVI	CVI
Stress ratio	0.1	0.1	0.1	0.1	0.1	0.1
Loading frequency/(Hz)	10	10	10	10	60	60
Fiber type	T-700™	T-700™	T-300™	T-300™	T-300™	T-300™
V_f	0.4	0.4	0.45	0.4	0.4	0.4
r_f/(μm)	3.5	3.5	3.5	3.5	3.5	3.5
Testing Temperature/(°C)	800	800	550	800	1300	1500
σ_{uts}/(MPa)	320	150	487	280	304	261
m	5	5	5	5	5	5

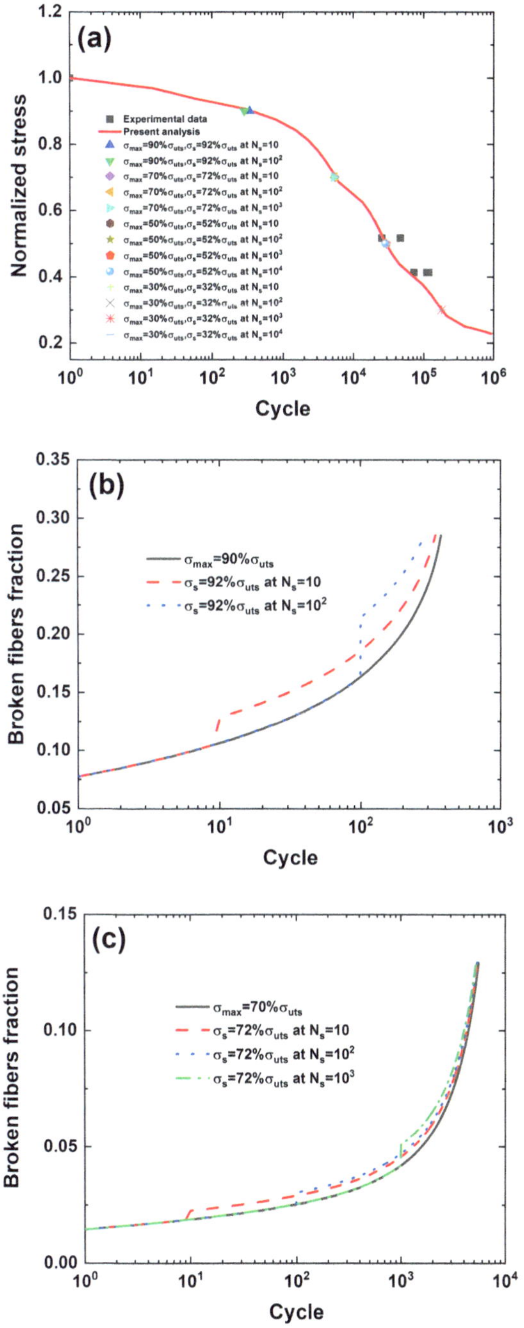

Fig. 4.2 Unidirectional C/SiC composite at T = 800 °C in air atmosphere for **a** the experimental and predicted fatigue life S–N curves under stochastic overloading; **b** the broken fibers fraction versus applied cycle number under $\sigma_{max} = 90\%\sigma_{uts}$ and $\sigma_s = 92\%\sigma_{uts}$ at $N_s = 10$ and 10^2; **c** the broken fibers fraction versus applied cycle number under $\sigma_{max} = 70\%\sigma_{uts}$ and $\sigma_s = 72\%\sigma_{uts}$ at $N_s = 10, 10^2$, and 10^3; **d** the broken fibers fraction versus applied cycle number under $\sigma_{max} = 50\%\sigma_{uts}$ and $\sigma_s = 52\%\sigma_{uts}$ at $N_s = 10, 10^2, 10^3$, and 10^4; and **e** the broken fibers fraction versus applied cycle number under $\sigma_{max} = 30\%\sigma_{uts}$ and $\sigma_s = 32\%\sigma_{uts}$ at $N_s = 10, 10^2, 10^3$, and 10^4

Fig. 4.2 (continued)

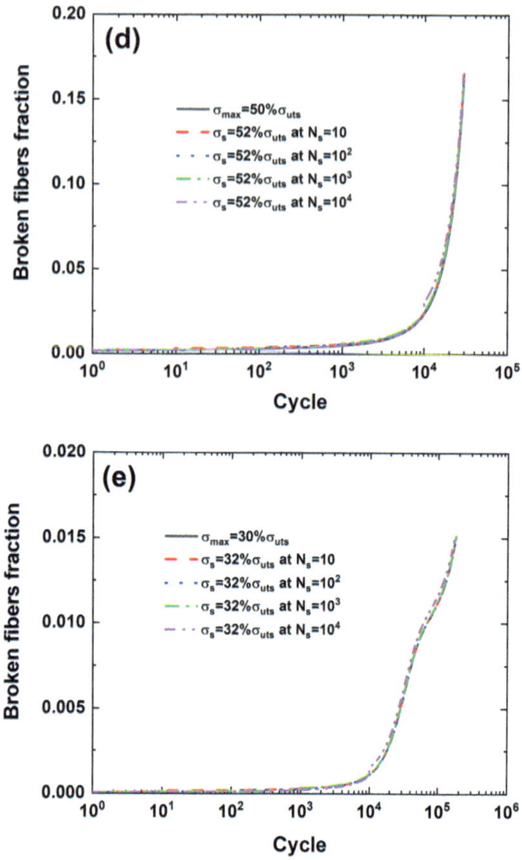

overloading stresses and occurrence applied cycle number. The fatigue life decreases when stochastic overloading stress occurs under constant fatigue peak stress loading. The fatigue life decreasing rate increases with stochastic overloading stress level and occurrence applied cycle number.

- Under fatigue peak stress $\sigma_{max} = 90\%\sigma_{uts}$ and stochastic overloading stress $\sigma_s = 92\%\sigma_{uts}$ at different occurrence applied cycle number $N_s = 10$ and 10^2, the fatigue life decreases from $N_f = 374$ to 341 and 285, and the fatigue life decreasing rate is $\Lambda = 8.8$ and 23.8%.
- Under fatigue peak stress $\sigma_{max} = 70\%\sigma_{uts}$ and stochastic overloading stress $\sigma_s = 72\%\sigma_{uts}$ at different occurrence applied cycle number $N_s = 10$, 10^2, and 10^3, the fatigue life decreases from $N_f = 5550$ to 5426, 5378, and 5243, and the fatigue life decreasing rate is $\Lambda = 2.2$, 3.1 and 5.5%.
- Under fatigue peak stress $\sigma_{max} = 50\%\sigma_{uts}$ and stochastic overloading stress $\sigma_s = 52\%\sigma_{uts}$ at different occurrence applied cycle number $N_s = 10$, 10^2, 10^3, and

10^4, the fatigue life decreases from $N_f = 29{,}300$ to 29,254, 29,239, 29,202, and 28,800, and the fatigue life decreasing rate is $\Lambda = 0.15$, 0.2, 0.3, and 1.7%.
- Under fatigue peak stress $\sigma_{max} = 30\%\sigma_{uts}$ and stochastic overloading stress $\sigma_s = 32\%\sigma_{uts}$ at occurrence applied cycle number $N_s = 10$, 10^2, 10^3, 10^4, and 10^5, the fatigue life decreases from $N_f = 180{,}719$ to 179,869, 179,579, 178,909, and 172,322, and the fatigue life decreasing rate is $\Lambda = 0.47$, 0.63, 1.0, and 4.6%.

Figure 4.2b–e and Table 4.2 show the predicted broken fibers fraction under different stochastic overloading stress levels and occurrence applied cycle number.

- Under fatigue peak stress $\sigma_{max} = 90\%\sigma_{uts}$ and stochastic overloading stress $\sigma_s = 92\%\sigma_{uts}$ at different occurrence applied cycle number $N_s = 10$ and 10^2, the broken fibers fraction increases from $P_f = 10.6$ to 12.8% at $N_s = 10$, and from $P_f = 16.3$ to 21.4% at $N_s = 10^2$.
- Under fatigue peak stress $\sigma_{max} = 70\%\sigma_{uts}$ and stochastic overloading stress $\sigma_s = 72\%\sigma_{uts}$ at different occurrence applied cycle number $N_s = 10$, 10^2, and 10^3, the broken fibers fraction increases from $P_f = 1.8$ to 2.2% at $N_s = 10$, from $P_f = 2.5$ to 3.0% at $N_s = 10^2$, and from $P_f = 4.2$ to 5.1% at $N_s = 10^3$.
- Under fatigue peak stress $\sigma_{max} = 50\%\sigma_{uts}$ and stochastic overloading stress $\sigma_s = 52\%\sigma_{uts}$ at different occurrence applied cycle number $N_s = 10$, 10^2, 10^3, and 10^4, the broken fibers fraction increases from $P_f = 0.24$ to 0.30% at $N_s = 10$, from $P_f = 0.32$ to 0.40% at $N_s = 10^2$, from $P_f = 0.51$ to 0.64% at $N_s = 10^3$, and from $P_f = 2.4$ to 3.1% at $N_s = 10^4$.
- Under fatigue peak stress $\sigma_{max} = 30\%\sigma_{uts}$ and stochastic overloading stress $\sigma_s = 32\%\sigma_{uts}$ at different occurrence applied cycle number $N_s = 10$, 10^2, 10^3, 10^4, and 10^5, the broken fibers fraction increases from $P_f = 1.11 \times 10^{-4}$ to 1.64×10^{-4} at $N_s = 10$, from $P_f = 0.01$ to 0.02% at $N_s = 10^2$, from $P_f = 0.02\%$ to 0.03% at $N_s = 10^3$, and from $P_f = 0.11$ to 0.15% at $N_s = 10^4$.

4.3.2 *Fatigue Life Prediction of Cross-ply C/SiC Composite Under Different Stochastic Overloading Stresses at T = 800 °C in Air Atmosphere*

Li (2015) performed experimental investigations on tension–tension fatigue behavior of cross-ply C/SiC composite at $T = 800$ °C in air atmosphere. The cross-ply T − 700™ carbon fiber-reinforced silicon carbide composite was fabricated using the HP method. The fibers volume fraction in the composite was approximately $V_f = 0.4$, and the monotonic tensile strength was approximately $\sigma_{uts} = 150$ MPa. The fatigue loading stress ratio was $R = 0.1$, and the fatigue loading frequency was $f = 10$ Hz. The fatigue limit stress was approximately $\sigma_{limit} = 26\%\sigma_{uts}$.

Figure 4.3a shows the experimental and predicted fatigue life S–N curves of cross-ply C/SiC composite at $T = 800$ °C in air atmosphere under different stochastic overloading stress levels and occurrence applied cycle number. The fatigue life of cross-ply C/SiC composite decreases when stochastic overloading stress occurs under

Table 4.2 Fatigue life and broken fibers fraction of unidirectional C/SiC composite under stochastic loading at $T = 800$ °C in air atmosphere

$\sigma_{max}/\sigma_{uts} = 90\%$	Cycle number	1	10	10^2	374			
	$P_f/(\%)$	7.7	10.6	16.3	28.5			
$\sigma_s/\sigma_{uts} = 92\%$ $N_s = 10$	Cycle number	1	10	10^2	341			
	$P_f/(\%)$	7.7	12.8	18.5	28.5			
$\sigma_s/\sigma_{uts} = 92\%$ $N_s = 10^2$	Cycle number	1	10	10^2	285			
	$P_f/(\%)$	7.7	10.6	21.4	28.5			
$\sigma_{max}/\sigma_{uts} = 70\%$	Cycle number	1	10	10^2	10^3	5550		
	$P_f/(\%)$	1.4	1.8	2.5	4.2	12.9		
$\sigma_s/\sigma_{uts} = 72\%$ $N_s = 10$	Cycle number	1	10	10^2	10^3	5426		
	$P_f/(\%)$	1.4	2.2	2.9	4.5	12.9		
$\sigma_s/\sigma_{uts} = 72\%$ $N_s = 10^2$	Cycle number	1	10	10^2	10^3	5378		
	$P_f/(\%)$	1.4	1.8	3.0	4.7	12.9		
$\sigma_s/\sigma_{uts} = 72\%$ $N_s = 10^3$	Cycle number	1	10	10^2	10^3	5243		
	$P_f/(\%)$	1.4	1.8	2.5	5.0	12.9		
$\sigma_{max}/\sigma_{uts} = 50\%$	Cycle number	1	10	10^2	10^3	10^4	29,300	
	$P_f/(\%)$	0.19	0.24	0.32	0.51	2.4	16.5	
$\sigma_s/\sigma_{uts} = 52\%$ $N_s = 10$	Cycle number	1	10	10^2	10^3	10^4	29,254	
	$P_f/(\%)$	0.19	0.30	0.38	0.57	2.4	16.5	
$\sigma_s/\sigma_{uts} = 52\%$ $N_s = 10^2$	Cycle number	1	10	10^2	10^3	10^4	29,239	
	$P_f/(\%)$	0.19	0.24	0.40	0.59	2.5	16.5	
$\sigma_s/\sigma_{uts} = 52\%$ $N_s = 10^3$	Cycle number	1	10	10^2	10^3	10^4	29,202	
	$P_f/(\%)$	0.19	0.24	0.32	0.64	2.5	16.5	
$\sigma_s/\sigma_{uts} = 52\%$ $N_s = 10^4$	Cycle number	1	10	10^2	10^3	10^4	28,800	
	$P_f/(\%)$	0.19	0.24	0.32	0.51	3.1	16.5	
$\sigma_{max}/\sigma_{uts} = 30\%$	Cycle number	1	10	10^2	10^3	10^4	10^5	180,719
	$P_f/(\%)$	0.009	0.01	0.015	0.02	0.11	1.1	0.01514
$\sigma_s/\sigma_{uts} = 32\%$ $N_s = 10$	Cycle number	1	10	10^2	10^3	10^4	10^5	179,869
	$P_f/(\%)$	0.009	0.02	0.02	0.03	0.11	1.12	1.5
$\sigma_s/\sigma_{uts} = 32\%$ $N_s = 10^2$	Cycle number	1	10	10^2	10^3	10^4	10^5	179,579
	$P_f/(\%)$	0.009	0.01	0.02	0.03	0.11	1.1	1.5
$\sigma_s/\sigma_{uts} = 32\%$ $N_s = 10^3$	Cycle number	1	10	10^2	10^3	10^4	10^5	178,909
	$P_f/(\%)$	0.009	0.01	0.015	0.03	0.12	1.13	1.51
$\sigma_s/\sigma_{uts} = 0.32$ $N_s = 10^4$	Cycle number	1	10	10^2	10^3	10^4	10^5	172,322
	$P_f/(\%)$	0.009	0.01	0.015	0.02	0.16	1.2	0.01514

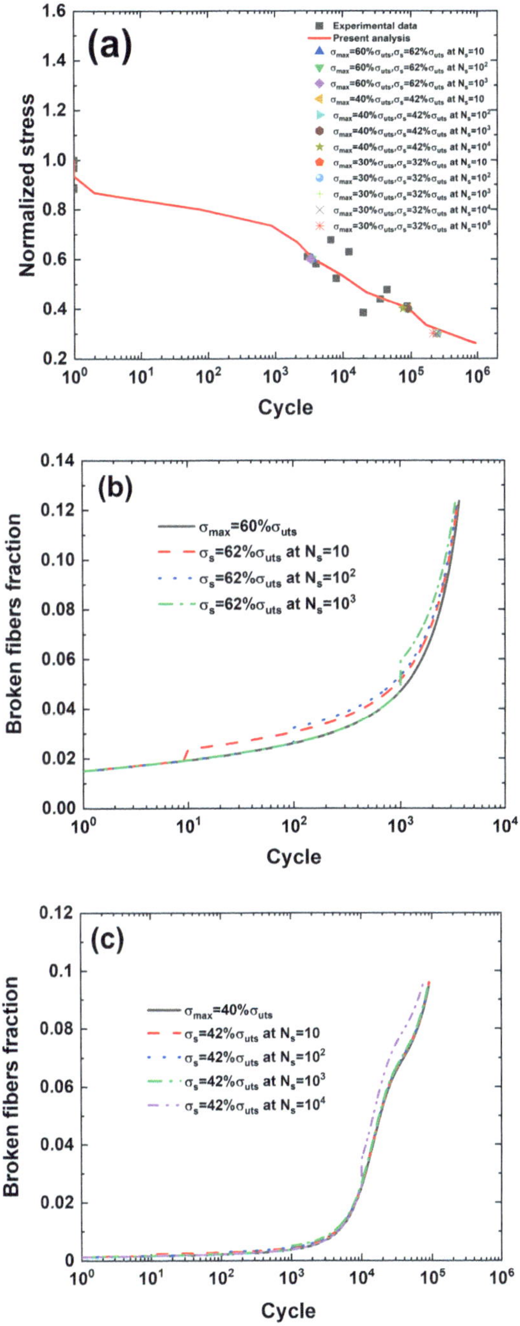

Fig. 4.3 Cross-ply C/SiC composite at $T = 800$ °C in air atmosphere for **a** the experimental and predicted fatigue life S–N curves under stochastic overloading; **b** the broken fibers fraction versus applied cycle number under $\sigma_{max} = 60\%\sigma_{uts}$ and $\sigma_s = 62\%\sigma_{uts}$ at $N_s = 10$, 10^2, and 10^3; **c** the broken fibers fraction versus applied cycle number under $\sigma_{max} = 40\%\sigma_{uts}$ and $\sigma_s = 42\%\sigma_{uts}$ at $N_s = 10$, 10^2, 10^3, and 10^4; and **d** the broken fibers fraction versus applied cycle number under $\sigma_{max} = 30\%\sigma_{uts}$ and $\sigma_s = 32\%\sigma_{uts}$ at $N_s = 10$, 10^2, 10^3, 10^4, and 10^5

Fig. 4.3 (continued)

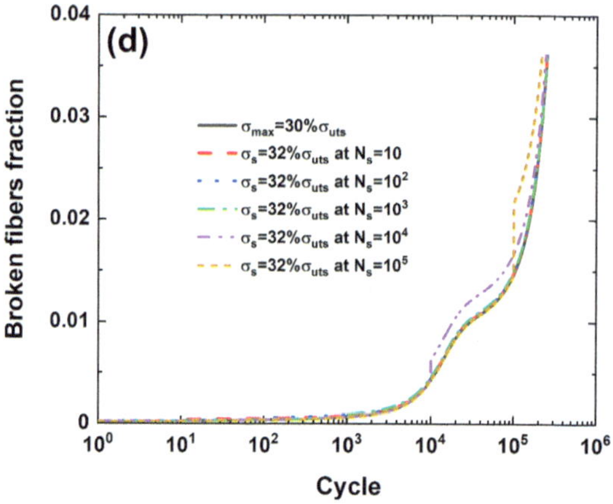

constant fatigue peak stress loading. The fatigue life decreasing rate increases with stochastic overloading stress level and occurrence applied cycle number.

- Under fatigue peak stress $\sigma_{\max} = 60\%\sigma_{\text{uts}}$ and stochastic overloading stress $\sigma_{\text{s}} = 62\%\sigma_{\text{uts}}$ at different occurrence applied cycle number $N_{\text{s}} = 10$, 10^2, and 10^3, the fatigue life decreases from $N_{\text{f}} = 3643$ to 3538, 3496, and 3347, and the fatigue life decreasing rate is $\Lambda = 2.8$, 4.0, and 8.1%.
- Under fatigue peak stress $\sigma_{\max} = 40\%\sigma_{\text{uts}}$ and stochastic overloading stress $\sigma_{\text{s}} = 42\%\sigma_{\text{uts}}$ at different occurrence applied cycle number $N_{\text{s}} = 10$, 10^2, 10^3, and 10^4, the fatigue life decreases from $N_{\text{f}} = 91{,}494$ to 90,632, 90,339, 89,487, and 75,354, and the fatigue life decreasing rate is $\Lambda = 0.9$, 1.2, 2.2, and 17.6%.
- Under fatigue peak stress $\sigma_{\max} = 30\%\sigma_{\text{uts}}$ and stochastic overloading stress $\sigma_{\text{s}} = 32\%\sigma_{\text{uts}}$ at different occurrence applied cycle number $N_{\text{s}} = 10$, 10^2, 10^3, 10^4, and 10^5, the fatigue life decreases from $N_{\text{f}} = 248{,}232$ to 247,635, 247,430, 246,860, 238,968, and 213,538, and the fatigue life decreasing rate is $\Lambda = 0.24$, 0.32, 0.55, 3.7, and 13.9%.

Figure 4.3b–d and Table 4.3 show the predicted broken fibers fraction under different stochastic overloading stress levels and occurrence applied cycle number. The broken fibers fraction increases when stochastic overloading stress occurs. The difference of broken fibers fraction between original fatigue peak stress and stochastic overloading stress increases with occurrence applied cycle number.

- Under fatigue peak stress $\sigma_{\max} = 60\%\sigma_{\text{uts}}$ and stochastic overloading stress $\sigma_{\text{s}} = 62\%\sigma_{\text{uts}}$ at different occurrence applied cycle number $N_{\text{s}} = 10$, 10^2, and 10^3, the broken fibers fraction increases from $P_{\text{f}} = 1.9$ to 2.3% at $N_{\text{s}} = 10$, from $P_{\text{f}} = 2.6$ to 3.2% at $N_{\text{s}} = 10^2$, and from $P_{\text{f}} = 4.7$ to 5.9% at $N_{\text{s}} = 10^3$.
- Under fatigue peak stress $\sigma_{\max} = 40\%\sigma_{\text{uts}}$ and stochastic overloading stress $\sigma_{\text{s}} = 42\%\sigma_{\text{uts}}$ at different occurrence applied cycle number $N_{\text{s}} = 10$, 10^2, 10^3, and 10^4, the broken fibers fraction increases from $P_{\text{f}} = 0.16$ to 0.21% at $N_{\text{s}} = 10$,

Table 4.3 Fatigue life and broken fibers fraction of cross-ply C/SiC composite under stochastic loading at $T = 800$ °C in air atmosphere

$\sigma_{max}/\sigma_{uts} = 60\%$	Cycle number	1	10	10^2	10^3	3643		
	$P_f/(\%)$	1.49	1.92	2.61	4.72	12.3		
$\sigma_s/\sigma_{uts} = 62\%$ $N_s = 10$	Cycle number	1	10	10^2	10^3	3538		
	$P_f/(\%)$	1.49	2.37	3.05	5.16	12.3		
$\sigma_s/\sigma_{uts} = 62\%$ $N_s = 10^2$	Cycle number	1	10	10^2	10^3	3496		
	$P_f/(\%)$	1.49	1.92	3.22	5.33	12.3		
$\sigma_s/\sigma_{uts} = 62\%$ $N_s = 10^3$	Cycle number	1	10	10^2	10^3	3347		
	$P_f/(\%)$	1.49	1.92	2.61	5.92	12.3		
$\sigma_{max}/\sigma_{uts} = 40\%$	Cycle number	1	10	10^2	10^3	10^4	91,494	
	$P_f/(\%)$	0.13	0.16	0.22	0.37	2.53	9.59	
$\sigma_s/\sigma_{uts} = 42\%$ $N_s = 10$	Cycle number	1	10	10^2	10^3	10^4	90,632	
	$P_f/(\%)$	0.13	0.22	0.27	0.43	2.58	9.59	
$\sigma_s/\sigma_{uts} = 42\%$ $N_s = 10^2$	Cycle number	1	10	10^2	10^3	10^4	90,339	
	$P_f/(\%)$	0.13	0.16	0.29	0.45	2.60	9.59	
$\sigma_s/\sigma_{uts} = 42\%$ $N_s = 10^3$	Cycle number	1	10	10^2	10^3	10^4	89,487	
	$P_f/(\%)$	0.13	0.16	0.22	0.49	2.66	9.59	
$\sigma_s/\sigma_{uts} = 42\%$ $N_s = 10^4$	Cycle number	1	10	10^2	10^3	10^4	75,354	
	$P_f/(\%)$	0.13	0.16	0.22	0.37	3.47	9.59	
$\sigma_{max}/\sigma_{uts} = 30\%$	Cycle number	1	10	10^2	10^3	10^4	10^5	248,232
	$P_f/(\%)$	0.022	0.028	0.038	0.066	0.43	1.46	3.61
$\sigma_s/\sigma_{uts} = 32\%$ $N_s = 10$	Cycle number	1	10	10^2	10^3	10^4	10^5	247,635
	$P_f/(\%)$	0.022	0.042	0.052	0.079	0.44	1.47	3.61
$\sigma_s/\sigma_{uts} = 32\%$ $N_s = 10^2$	Cycle number	1	10	10^2	10^3	10^4	10^5	247,430
	$P_f/(\%)$	0.022	0.028	0.056	0.084	0.45	1.48	3.61
$\sigma_s/\sigma_{uts} = 32\%$ $N_s = 10^3$	Cycle number	1	10	10^2	10^3	10^4	10^5	246,860
	$P_f/(\%)$	0.022	0.029	0.038	0.097	0.46	1.49	3.61
$\sigma_s/\sigma_{uts} = 32\%$ $N_s = 10^4$	Cycle number	1	10	10^2	10^3	10^4	28,800	238,968
	$P_f/(\%)$	0.022	0.029	0.038	0.066	0.63	1.67	3.61
$\sigma_s/\sigma_{uts} = 32\%$ $N_s = 10^5$	Cycle number	1	10	10^2	10^3	10^4	28,800	213,538
	$P_f/(\%)$	0.022	0.029	0.038	0.066	0.43	2.17	3.61

from $P_f = 0.22$ to 0.29% at $N_s = 10^2$, from $P_f = 0.37$ to 0.49% at $N_s = 10^3$, and from $P_f = 2.5$ to 3.4% at $N_s = 10^4$.

- Under fatigue peak stress $\sigma_{max} = 30\%\sigma_{uts}$ and stochastic overloading stress $\sigma_s = 32\%\sigma_{uts}$ at different occurrence applied cycle number $N_s = 10$, 10^2, 10^3, 10^4, and 10^5, the broken fibers fraction increases from $P_f = 0.02$ to 0.04% at $N_s = 10$, from $P_f = 0.03$ to 0.05% at $N_s = 10^2$, from $P_f = 0.06$ to 0.09% at $N_s = 10^3$, from $P_f = 0.42$ to 0.63% at $N_s = 10^4$, and from $P_f = 1.46$ to 2.17% at $N_s = 10^5$.

4.3.3 Fatigue Life Prediction of 2D Woven C/SiC Composite Under Different Stochastic Overloading Stresses at $T =$ 550 °C in Air Atmosphere

Mall and Engesser (2006) performed experimental investigations on tension–tension fatigue behavior of 2D woven C/SiC composite at T = 550 °C in air atmosphere. The 2D T − 300™ carbon fiber-reinforced silicon carbide composite was fabricated using the chemical vapor infiltrated (CVI) method. The fibers volume fraction in the composite was approximately $V_f = 0.45$, and the tensile strength was approximately σ_{uts} = 487 MPa. The fatigue loading stress ratio was R = 0.1, and the fatigue loading frequency was f = 10 Hz. The fatigue limit stress was approximately σ_{limit} = 30%σ_{uts}.

Figure 4.4a shows the experimental and predicted fatigue life S–N curves of 2D C/SiC composite at T = 550 °C in air atmosphere under different stochastic overloading stresses and occurrence applied cycle number. The fatigue life of 2D C/SiC composite decreases when stochastic overloading stress occurs under cyclic fatigue loading. The fatigue life decreasing rate increases with stochastic overloading stress level and occurrence applied cycle number.

- Under fatigue peak stress σ_{max} = 70%σ_{uts} and stochastic overloading stress σ_s = 72%σ_{uts} at different occurrence applied cycle number N_s = 10, 10^2, and 10^3, the fatigue life decreases from N_f = 6095 to 5980, 5895, and 5624, and the fatigue life decreasing rate is Λ = 1.8, 3.2, and 7.7%.
- Under fatigue peak stress σ_{max} = 60%σ_{uts} and stochastic overloading stress σ_s = 62%σ_{uts} at different occurrence applied cycle number N_s = 10, 10^2, 10^3, and 10^4, the fatigue life decreases from N_f = 19,903 to 19,703, 19,592, 19,365, and 18,212, and the fatigue life decreasing rate is Λ = 1.0, 1.5, 2.7, and 8.5%.
- Under fatigue peak stress σ_{max} = 40%σ_{uts} and stochastic overloading stress σ_s = 42%σ_{uts} at different occurrence applied cycle number N_s = 10, 10^2, 10^3, and 10^4, the fatigue life decreases from N_f = 114,447 to 114,403, 114,381, 114,345, and 114,234, and the fatigue life decreasing rate is Λ = 0.03, 0.05, 0.08, and 0.1%.

Figure 4.4b–d and Table 4.4 show the predicted broken fibers fraction under different stochastic overloading stresses and occurrence applied cycle number. The broken fibers fraction increases when stochastic overloading stress occurs. The difference of broken fibers fraction between original fatigue peak stress and stochastic overloading stress increases with occurrence applied cycle number.

- Under fatigue peak stress σ_{max} = 70%σ_{uts} and stochastic overloading stress σ_s = 72%σ_{uts} at different occurrence applied cycle number N_s = 10, 10^2, and 10^3, the broken fibers fraction increases from P_f = 3.5 to 4.3% at N_s = 10, from P_f = 5.5 to 6.7% at N_s = 10^2, and from P_f = 9.7 to 12.3% at N_s = 10^3.
- Under fatigue peak stress σ_{max} = 60%σ_{uts} and stochastic overloading stress σ_s = 62%σ_{uts} at different occurrence applied cycle number N_s = 10, 10^2, 10^3, and 10^4, the broken fibers fraction increases from P_f = 1.3 to 1.6% at N_s = 10, from

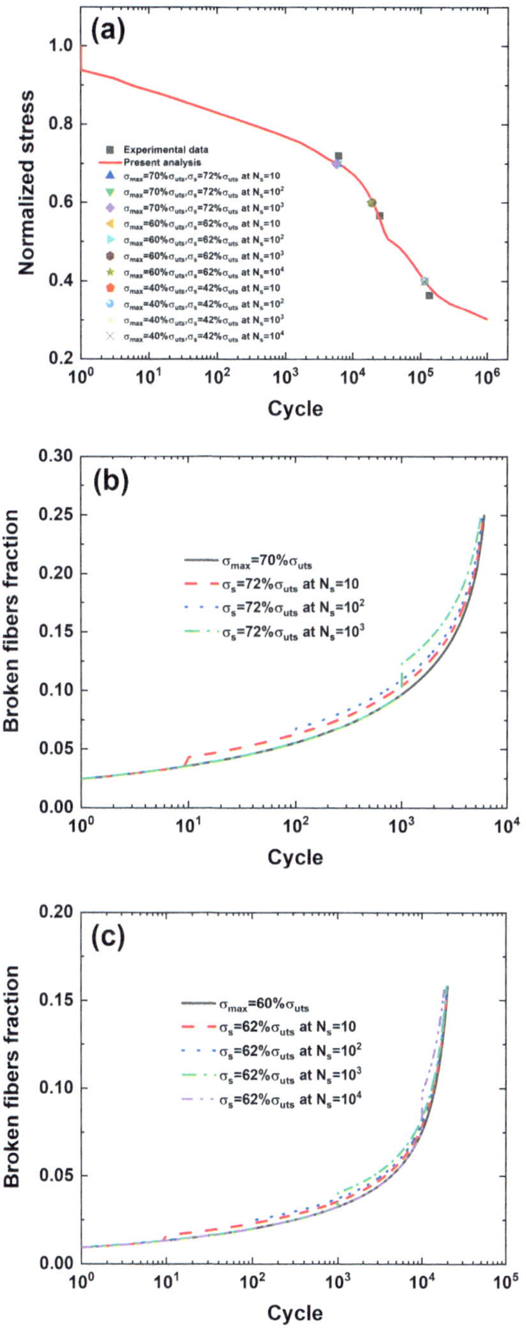

Fig. 4.4 2D woven C/SiC composite at $T = 550$ °C in air atmosphere for **a** the experimental and predicted fatigue life S–N curves under stochastic overloading; **b** the broken fibers fraction versus applied cycle number under $\sigma_{max} = 70\%\sigma_{uts}$ and $\sigma_s = 72\%\sigma_{uts}$ at $N_s = 10$, 10^2, and 10^3; **c** the broken fibers fraction versus applied cycle number under $\sigma_{max} = 60\%\sigma_{uts}$ and $\sigma_s = 62\%\sigma_{uts}$ at $N_s = 10$, 10^2, 10^3, and 10^4; and **d** the broken fibers fraction versus applied cycle number under $\sigma_{max} = 40\%\sigma_{uts}$ and $\sigma_s = 42\%\sigma_{uts}$ at $N_s = 10$, 10^2, 10^3, and 10^4

Fig. 4.4 (continued)

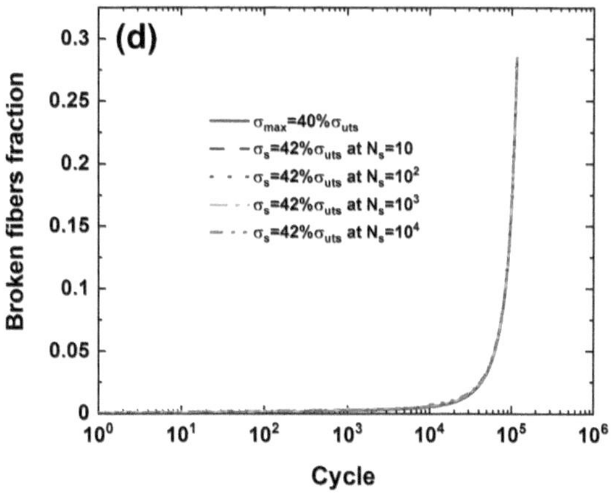

$P_f = 2.0$ to 2.4% at $N_s = 10^2$, from $P_f = 3.2$ to 4.0% at $N_s = 10^3$, and from $P_f = 7.5$ to 9.8% at $N_s = 10^4$.

- Under fatigue peak stress $\sigma_{max} = 40\%\sigma_{uts}$ and stochastic overloading stress $\sigma_s = 42\%\sigma_{uts}$ at different occurrence applied cycle number $N_s = 10$, 10^2, 10^3, and 10^4, the broken fibers fraction increases from $P_f = 0.11$ to 0.15% at $N_s = 10$, from $P_f = 0.17$ to 0.23% at $N_s = 10^2$, from $P_f = 0.27$ to 0.35% at $N_s = 10^3$, and from $P_f = 0.57$ to 0.76% at $N_s = 10^4$.

4.3.4 Fatigue Life Prediction of 2.5D Woven C/SiC Composite Under Different Stochastic Overloading Stresses at $T = 800$ °C in Air Atmosphere

Yang (2011) performed the experimental investigations on tension–tension fatigue behavior of 2.5D woven C/SiC composite at $T = 800$ °C in air atmosphere. The 2.5D $T - 300^{TM}$ carbon fiber-reinforced silicon carbide composite was fabricated using CVI method. The fibers volume fraction in the composite was approximately $V_f = 0.4$, and the tensile strength was approximately $\sigma_{uts} = 280$ MPa. The fatigue loading stress ratio was $R = 0.1$, and the fatigue loading frequency was $f = 10$ Hz. The fatigue limit stress was approximately $\sigma_{limit} = 28\%\sigma_{uts}$.

Figure 4.5a shows the experimental and predicted fatigue life S–N curves of 2.5D C/SiC composite at $T = 800$ °C in air atmosphere under different stochastic overloading stresses and occurrence applied cycle number. The fatigue life of 2.5D C/SiC decreases when stochastic overloading stress occurs under cyclic fatigue loading. The fatigue life decreasing rate increases with stochastic overloading stress level and occurrence applied cycle number.

- Under fatigue peak stress $\sigma_{max} = 80\%\sigma_{uts}$ and stochastic overloading stress $\sigma_s = 82\%\sigma_{uts}$ at different occurrence applied cycle number $N_s = 10$ and 10^2, the fatigue

Table 4.4 Fatigue life and broken fibers fraction of 2D woven C/SiC composite under stochastic loading at $T = 550$ °C in air atmosphere

$\sigma_{max}/\sigma_{uts} = 70\%$	Cycle number	1	10	10^2	10^3	6095		
	$P_f/(\%)$	2.42	3.57	5.55	9.71	24.9		
$\sigma_s/\sigma_{uts} = 72\%$ $N_s = 10$	Cycle number	1	10	10^2	10^3	5980		
	$P_f/(\%)$	2.42	4.31	6.29	10.44	24.9		
$\sigma_s/\sigma_{uts} = 72\%$ $N_s = 10^2$	Cycle number	1	10	10^2	10^3	5895		
	$P_f/(\%)$	2.42	3.57	6.78	10.94	24.9		
$\sigma_s/\sigma_{uts} = 72\%$ $N_s = 10^3$	Cycle number	1	10	10^2	10^3	5624		
	$P_f/(\%)$	2.42	3.57	5.55	12.32	24.9		
$\sigma_{max}/\sigma_{uts} = 60\%$	Cycle number	1	10	10^2	10^3	10^4	19,903	
	$P_f/(\%)$	0.93	1.34	2.01	3.26	7.56	15.82	
$\sigma_s/\sigma_{uts} = 62\%$ $N_s = 10$	Cycle number	1	10	10^2	10^3	10^4	19,703	
	$P_f/(\%)$	0.93	1.64	2.32	3.56	7.86	15.8	
$\sigma_s/\sigma_{uts} = 62\%$ $N_s = 10^2$	Cycle number	1	10	10^2	10^3	10^4	19,592	
	$P_f/(\%)$	0.93	1.34	2.48	3.73	8.03	15.8	
$\sigma_s/\sigma_{uts} = 62\%$ $N_s = 10^3$	Cycle number	1	10	10^2	10^3	10^4	19,365	
	$P_f/(\%)$	0.93	1.34	2.01	4.05	8.35	15.8	
$\sigma_s/\sigma_{uts} = 62\%$ $N_s = 10^4$	Cycle number	1	10	10^2	10^3	10^4	18,212	
	$P_f/(\%)$	0.93	1.34	2.01	3.26	9.79	15.8	
$\sigma_{max}/\sigma_{uts} = 40\%$	Cycle number	1	10	10^2	10^3	10^4	10^5	114,447
	$P_f/(\%)$	0.079	0.114	0.169	0.267	0.57	17.2	28.6
$\sigma_s/\sigma_{uts} = 42\%$ $N_s = 10$	Cycle number	1	10	10^2	10^3	10^4	10^5	114,403
	$P_f/(\%)$	0.079	0.153	0.208	0.305	0.609	17.2	28.6
$\sigma_s/\sigma_{uts} = 42\%$ $N_s = 10^2$	Cycle number	1	10	10^2	10^3	10^4	10^5	114,381
	$P_f/(\%)$	0.079	0.114	0.227	0.324	0.628	17.2	28.6
$\sigma_s/\sigma_{uts} = 42\%$ $N_s = 10^3$	Cycle number	1	10	10^2	10^3	10^4	10^5	114,345
	$P_f/(\%)$	0.079	0.114	0.169	0.358	0.661	17.3	28.6
$\sigma_s/\sigma_{uts} = 42\%$ $N_s = 10^4$	Cycle number	1	10	10^2	10^3	10^4	10^5	114,234
	$P_f/(\%)$	0.079	0.114	0.169	0.267	0.763	17.4	28.6

life decreases from $N_f = 736$ to 719 and 698, and the fatigue life decreasing rate is $\Lambda = 2.3$ and 5.1%.

- Under fatigue peak stress $\sigma_{max} = 60\%\sigma_{uts}$ and stochastic overloading stress $\sigma_s = 62\%\sigma_{uts}$ at different occurrence applied cycle number $N_s = 10$, 10^2, and 10^3, the fatigue life decreases from $N_f = 8314$ to 8221, 8168, and 8008, and the fatigue life decreasing rate is $\Lambda = 1.1$, 1.7, and 3.6%.
- Under fatigue peak stress $\sigma_{max} = 50\%\sigma_{uts}$ and stochastic overloading stress $\sigma_s = 52\%\sigma_{uts}$ at different occurrence applied cycle number $N_s = 10$, 10^2, 10^3, and

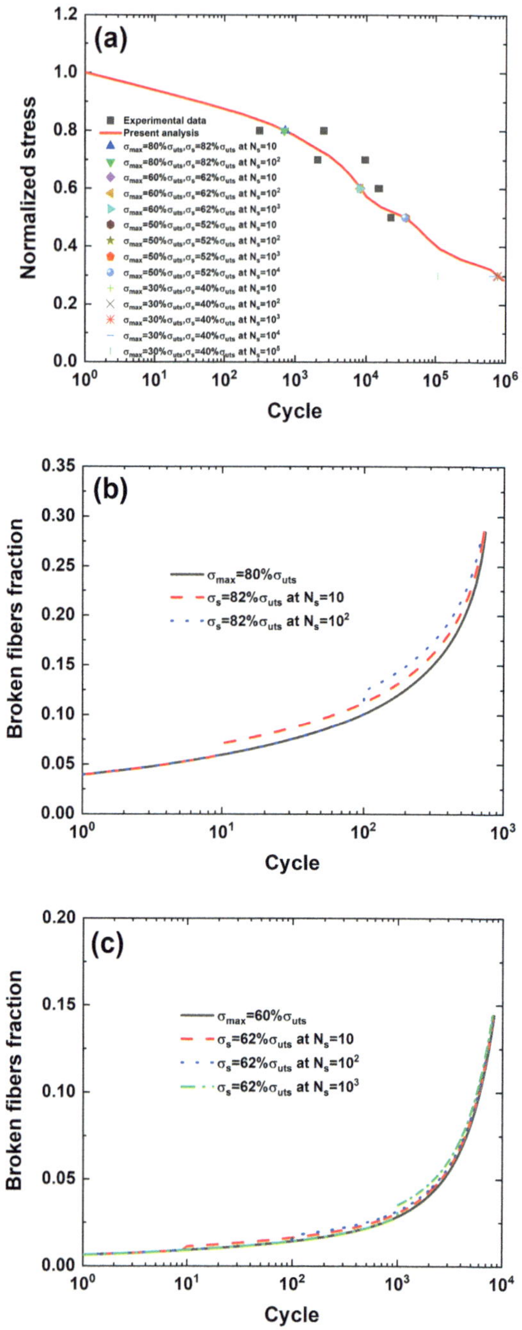

Fig. 4.5 2.5D woven C/SiC composite at $T = 800$ °C in air atmosphere for **a** the experimental and predicted fatigue life S–N curves under stochastic overloading; **b** the broken fibers fraction versus applied cycle number under $\sigma_{max} = 80\%\sigma_{uts}$ and $\sigma_s = 82\%\sigma_{uts}$ at $N_s = 10$ and 10^2; **c** the broken fibers fraction versus applied cycle number under $\sigma_{max} = 60\%\sigma_{uts}$ and $\sigma_s = 62\%\sigma_{uts}$ at $N_s = 10$, 10^2, and 10^3; and **d** the broken fibers fraction versus applied cycle number under $\sigma_{max} = 30\%\sigma_{uts}$ and $\sigma_s = 40\%\sigma_{uts}$ at $N_s = 10$, 10^2, 10^3, 10^4, and 10^5

Fig. 4.5 (continued)

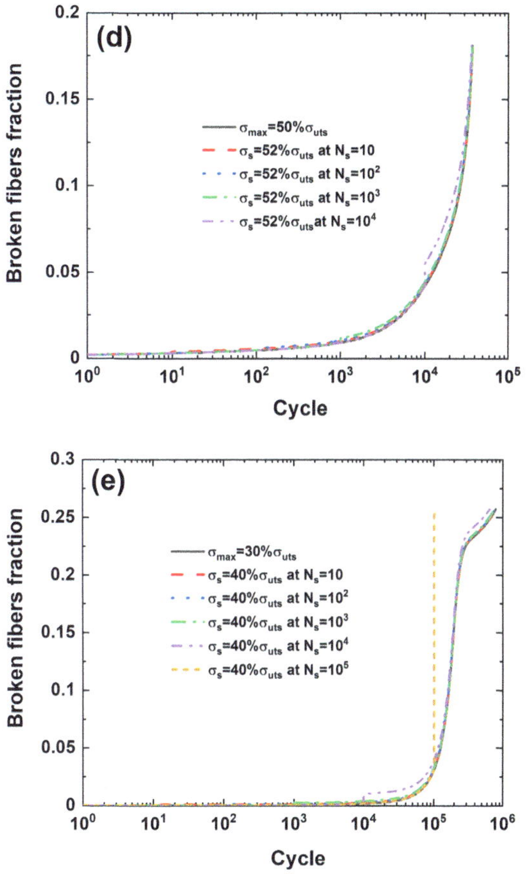

10^4, the fatigue life decreases from $N_f = 37{,}308$ to 37,254, 37,225, 37,145, and 36,374, and the fatigue life decreasing rate is $\Lambda = 0.14$, 0.22, 0.4, and 2.5%.

- Under fatigue peak stress $\sigma_{max} = 30\%\sigma_{uts}$ and stochastic overloading stress $\sigma_s = 40\%\sigma_{uts}$ at different occurrence applied cycle number $N_s = 10$, 10^2, 10^3, 10^4, and 10^5, the fatigue life decreases from $N_f = 790{,}756$ to 780,095, 774,501, 759,330, 650,314, and 107,300, and the fatigue life decreasing rate is $\Lambda = 1.3$, 2.0, 3.9, 17.7, and 86.4%.

Figure 4.5b–e and Table 4.5 show the predicted broken fibers fraction under different stochastic overloading stress and occurrence applied cycle number. The broken fibers fraction increases when stochastic overloading stress occurs. The difference of broken fibers fraction between original fatigue peak stress and stochastic overloading stress increases with occurrence applied cycle number.

- Under fatigue peak stress $\sigma_{max} = 80\%\sigma_{uts}$ and stochastic overloading stress $\sigma_s = 82\%\sigma_{uts}$ at different occurrence applied cycle number $N_s = 10$ and 10^2, the

Table 4.5 Fatigue life and broken fibers fraction of 2.5D woven C/SiC composite under stochastic loading at T = 800 °C in air atmosphere

$\sigma_{max}/\sigma_{uts}$ = 80%	Cycle number	1	10	10^2	736			
	P_f/(%)	3.93	5.94	10.1	28.5			
σ_s/σ_{uts} = 82% N_s = 10	Cycle number	1	10	10^2	719			
	P_f/(%)	3.93	7.1	11.2	28.5			
σ_s/σ_{uts} = 82% N_s = 10^2	Cycle number	1	10	10^2	698			
	P_f/(%)	3.93	5.94	12.4	28.5			
$\sigma_{max}/\sigma_{uts}$ = 60%	Cycle number	1	10	10^2	10^3	8314		
	P_f/(%)	0.64	0.93	1.44	2.85	14.4		
σ_s/σ_{uts} = 62% N_s = 10	Cycle number	1	10	10^2	10^3	8221		
	P_f/(%)	0.64	1.14	1.64	3.05	14.4		
σ_s/σ_{uts} = 62% N_s = 10^2	Cycle number	1	10	10^2	10^3	8168		
	P_f/(%)	0.64	0.93	1.76	3.17	14.4		
σ_s/σ_{uts} = 62% N_s = 10^3	Cycle number	1	10	10^2	10^3	8008		
	P_f/(%)	0.64	0.93	1.44	3.52	14.4		
$\sigma_{max}/\sigma_{uts}$ = 50%	Cycle number	1	10	10^2	10^3	10^4	37,308	
	P_f/(%)	0.21	0.31	0.47	0.91	4.21	18.1	
σ_s/σ_{uts} = 52% N_s = 10	Cycle number	1	10	10^2	10^3	10^4	37,254	
	P_f/(%)	0.21	0.39	0.55	0.99	4.29	18.1	
σ_s/σ_{uts} = 52% N_s = 10^2	Cycle number	1	10	10^2	10^3	10^4	37,225	
	P_f/(%)	0.21	0.31	0.59	1.03	4.33	18.1	
σ_s/σ_{uts} = 52% N_s = 10^3	Cycle number	1	10	10^2	10^3	10^4	37,145	
	P_f/(%)	0.21	0.31	0.47	1.16	4.45	18.1	
σ_s/σ_{uts} = 52% N_s = 10^4	Cycle number	1	10	10^2	10^3	10^4	36,374	
	P_f/(%)	0.21	0.31	0.47	0.91	5.50	18.1	
$\sigma_{max}/\sigma_{uts}$ = 30%	Cycle number	1	10	10^2	10^3	10^4	10^5	790,756
	P_f/(%)	0.0099	0.014	0.022	0.042	0.185	3.08	25.7
σ_s/σ_{uts} = 40% N_s = 10	Cycle number	1	10	10^2	10^3	10^4	10^5	780,095
	P_f/(%)	0.0099	0.08	0.087	0.11	0.25	3.14	25.7
σ_s/σ_{uts} = 40% N_s = 10^2	Cycle number	1	10	10^2	10^3	10^4	10^5	774,501
	P_f/(%)	0.0099	0.014	0.12	0.14	0.28	3.18	25.7
σ_s/σ_{uts} = 40% N_s = 10^3	Cycle number	1	10	10^2	10^3	10^4	10^5	769,330
	P_f/(%)	0.0099	0.014	0.022	0.24	0.38	3.27	25.7
σ_s/σ_{uts} = 40% N_s = 10^4	Cycle number	1	10	10^2	10^3	10^4	10^5	650,314
	P_f/(%)	0.0099	0.014	0.022	0.042	1.03	3.92	25.7
σ_s/σ_{uts} = 40% N_s = 10^4	Cycle number	1	10	10^2	10^3	10^4	10^5	107,300
	P_f/(%)	0.0099	0.014	0.022	0.042	0.19	25.2	25.7

broken fibers fraction increases from $P_f = 5.9$ to 7.1% at $N_s = 10$ and from $P_f =$ 10.0 to 12.4% at $N_s = 10^2$.

- Under fatigue peak stress $\sigma_{max} = 60\%\sigma_{uts}$ and stochastic overloading stress $\sigma_s = 62\%\sigma_{uts}$ at different occurrence applied cycle number $N_s = 10$, 10^2, and 10^3, the broken fibers fraction increases from $P_f = 0.9$ to 1.14% at $N_s = 10$, from $P_f = 1.4$ to 1.7% at $N_s = 10^2$, and from $P_f = 2.8$ to 3.5% at $N_s = 10^3$.
- Under fatigue peak stress $\sigma_{max} = 50\%\sigma_{uts}$ and stochastic overloading stress $\sigma_s = 52\%\sigma_{uts}$ at different occurrence applied cycle number $N_s = 10$, 10^2, 10^3, and 10^4, the broken fibers fraction increases from $P_f = 0.31$ to 0.39 at $N_s = 10$, from $P_f = 0.47$ to 0.59 at $N_s = 10^2$, from $P_f = 0.91$ to 1.16 at $N_s = 10^3$, and from $P_f = 4.2$ to 5.5% at $N_s = 10^4$.
- Under fatigue peak stress $\sigma_{max} = 30\%\sigma_{uts}$ and stochastic overloading stress $\sigma_s = 40\%\sigma_{uts}$ at different occurrence applied cycle number $N_s = 10$, 10^2, 10^3, 10^4, and 10^5, the broken fibers fraction increases from $P_f = 0.01$ to 0.08% at $N_s = 10$, from $P_f = 0.02$ to 0.12% at $N_s = 10^2$, from $P_f = 0.04$ to 0.23% at $N_s = 10^3$, from $P_f = 0.185$ to 1.03% at $N_s = 10^4$, and from $P_f = 3.1$ to 25.2% at $N_s = 10^5$.

4.3.5 Fatigue Life Prediction of 3D Braided C/SiC Composite Under Different Stochastic Overloading Stresses at T = 1300 and 1500 °C in Vacuum Atmosphere

Du et al. (2002), Du and Qiao (2011) performed the experimental investigations on tension–tension fatigue behavior of 3D braided C/SiC composite at $T = 1300$ and 1500 °C in vacuum atmosphere. The 3D T − 300™ carbon fiber-reinforced silicon carbide composite was fabricated using the CVI method. The fibers volume fraction in the composite was approximately $V_f = 0.4$, and the tensile strength was approximately $\sigma_{uts} = 304$ MPa. The fatigue loading stress ratio was $R = 0.1$, and the fatigue loading frequency was $f = 60$ Hz. The fatigue limit stress was approximately $\sigma_{limit} = 93\%\sigma_{uts}$.

Figure 4.6a shows the experimental and predicted fatigue life S–N curves of 3D braided C/SiC composite at $T = 1300$ °C in vacuum atmosphere under different stochastic overloading stresses and occurrence applied cycle number. The fatigue life decreases when stochastic overloading stress occurs under cyclic fatigue loading. The fatigue life decreasing rate increases with stochastic overloading stress level and occurrence applied cycle number.

- Under fatigue peak stress $\sigma_{max} = 95\%\sigma_{uts}$ and stochastic overloading stress $\sigma_s = 96\%\sigma_{uts}$ at different occurrence applied cycle number $N_s = 10$, 10^2, 10^3, and 10^4, the fatigue life decreases from $N_f = 143{,}725$ to 110,896, 106,541, 98,635, and 69,844, and the fatigue life decreasing rate is $\Lambda = 22.8$, 25.8, 31.1, and 51.4%.
- Under fatigue peak stress $\sigma_{max} = 94\%\sigma_{uts}$ and stochastic overloading stress $\sigma_s = 95\%\sigma_{uts}$ at different occurrence applied cycle number $N_s = 10$, 10^2, 10^3, 10^4, and 10^5, the fatigue life decreases from $N_f = 354{,}099$ to 277,452, 267,135, 248,309,

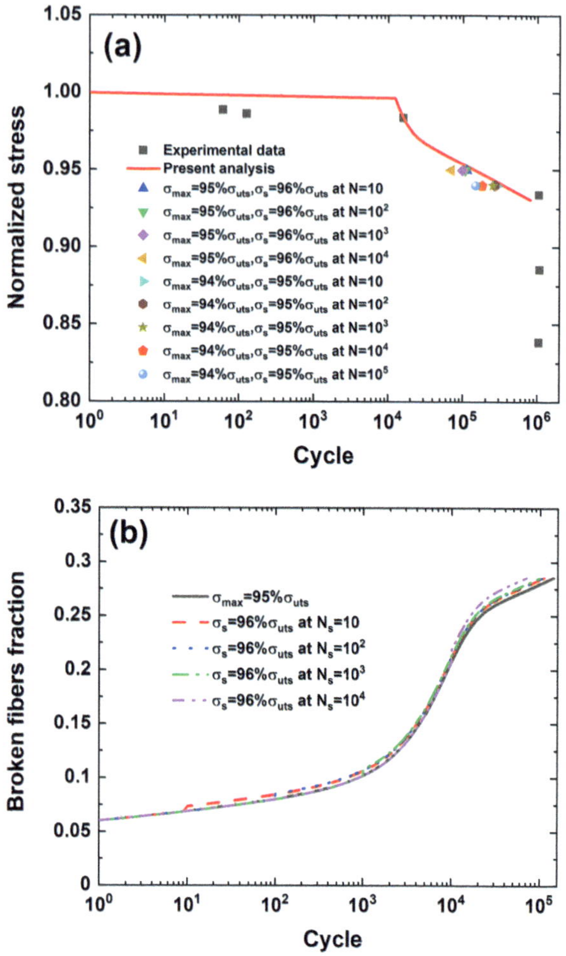

Fig. 4.6 3D braided C/SiC composite at $T = 1300$ °C for **a** the experimental and predicted fatigue life S–N curves under stochastic overloading; **b** the broken fibers fraction versus applied cycle number under $\sigma_{max} = 95\%\sigma_{uts}$ and $\sigma_s = 96\%\sigma_{uts}$ at $N_s = 10, 10^2, 10^3$, and 10^4; and **c** the broken fibers fraction versus applied cycle number under $\sigma_{max} = 94\%\sigma_{uts}$ and $\sigma_s = 95\%\sigma_{uts}$ at $N_s = 10, 10^2, 10^3, 10^4$, and 10^5

178,415, and 146,011, and the fatigue life decreasing rate is $\Lambda = 21.6, 24.5, 29.8, 49.6$, and 58.7%.

Figures 4.6b, c and Table 4.6 show the predicted broken fibers fraction under different stochastic overloading stresses and occurrence applied cycle number. The broken fibers fraction increases when stochastic overloading stress occurs. The difference of broken fibers fraction between original fatigue peak stress and stochastic overloading stress increases with occurrence applied cycle number.

- Under fatigue peak stress $\sigma_{max} = 95\%\sigma_{uts}$ and stochastic overloading stress $\sigma_s = 96\%\sigma_{uts}$ at different occurrence applied cycle number $N_s = 10, 10^2, 10^3$, and 10^4, the broken fibers fraction increases from $P_f = 6.9$ to 7.3% at $N_s = 10$, from $P_f = 8.0$ to 8.5% at $N_s = 10^2$, from $P_f = 10.2$ to 10.8% at $N_s = 10^3$, and from $P_f = 20.7$ to 21.9% at $N_s = 10^4$.

Table 4.6 Fatigue life and broken fibers fraction of 3D woven C/SiC composite under stochastic loading at $T = 1300$ °C

$\sigma_{max}/\sigma_{uts} = 95\%$	Cycle number	1	10	10^2	10^3	10^4	10^5	143,725
	$P_f/(\%)$	6.05	6.93	8.05	10.2	20.8	27.9	28.6
$\sigma_s/\sigma_{uts} = 96\%$ $N_s = 10$	Cycle number	1	10	10^2	10^3	10^4	10^5	110,896
	$P_f/(\%)$	6.05	7.37	8.48	10.6	21.2	28.4	28.6
$\sigma_s/\sigma_{uts} = 96\%$ $N_s = 10^2$	Cycle number	1	10	10^2	10^3	10^4	10^5	106,541
	$P_f/(\%)$	6.05	6.94	8.55	10.73	21.3	28.5	28.6
$\sigma_s/\sigma_{uts} = 96\%$ $N_s = 10^3$	Cycle number	1	10	10^2	10^3	10^4	98,635	
	$P_f/(\%)$	6.05	6.94	8.05	10.86	21.4	28.6	
$\sigma_s/\sigma_{uts} = 96\%$ $N_s = 10^4$	Cycle number	1	10	10^2	10^3	10^4	69,844	
	$P_f/(\%)$	6.05	6.94	8.05	10.23	21.9	28.6	
$\sigma_{max}/\sigma_{uts} = 94\%$	Cycle number	1	10	10^2	10^3	10^4	10^5	354,099
	$P_f/(\%)$	5.69	6.52	7.57	9.63	19.63	26.5	28.6
$\sigma_s/\sigma_{uts} = 95\%$ $N_s = 10$	Cycle number	1	10	10^2	10^3	10^4	10^5	277,452
	$P_f/(\%)$	5.69	6.94	7.99	10.04	20.05	26.9	28.6
$\sigma_s/\sigma_{uts} = 95\%$ $N_s = 10^2$	Cycle number	1	10	10^2	10^3	10^4	10^5	267,135
	$P_f/(\%)$	5.69	6.52	8.05	10.11	20.11	26.9	28.6
$\sigma_s/\sigma_{uts} = 95\%$ $N_s = 10^3$	Cycle number	1	10	10^2	10^3	10^4	10^5	248,309
	$P_f/(\%)$	5.69	6.52	7.57	10.23	20.23	27.09	28.6
$\sigma_s/\sigma_{uts} = 95\%$ $N_s = 10^4$	Cycle number	1	10	10^2	10^3	10^4	10^5	178,415
	$P_f/(\%)$	5.69	6.52	7.57	9.63	20.78	27.64	28.6
$\sigma_s/\sigma_{uts} = 95\%$ $N_s = 10^5$	Cycle number	1	10	10^2	10^3	10^4	10^5	146,011
	$P_f/(\%)$	5.69	6.52	7.57	9.63	19.63	27.97	28.6

- Under fatigue peak stress $\sigma_{max} = 94\%\sigma_{uts}$ and stochastic overloading stress $\sigma_s = 95\%\sigma_{uts}$ at $N_s = 10$, 10^2, 10^3, 10^4, and 10^5, the broken fibers fraction increases from $P_f = 6.5$ to 6.9% at $N_s = 10$, from $P_f = 7.5$ to 8.0% at $N_s = 10^2$, from $P_f = 10.0$ to 10.2% at $N_s = 10^3$, from $P_f = 19.6$ to 20.7% at $N_s = 10^4$, and from $P_f = 26.5$ to 27.9% at $N_s = 10^5$.

At $T = 1500$ °C in vacuum, the tensile strength was approximately $\sigma_{uts} = 261$ MPa. The fatigue loading stress ratio was $R = 0.1$, and the fatigue loading frequency was $f = 60$ Hz. The fatigue limit stress was approximately $\sigma_{limit} = 90\%\sigma_{uts}$.

Figure 4.7a shows the experimental and predicted fatigue life S–N curves of 3D C/SiC composite at $T = 1500$ °C in vacuum atmosphere under different stochastic overloading stresses and occurrence applied cycle number. The fatigue life decreases when stochastic overloading stress occurs under cyclic fatigue loading. The fatigue life decreasing rate increases with stochastic overloading stress level and occurrence applied cycle number.

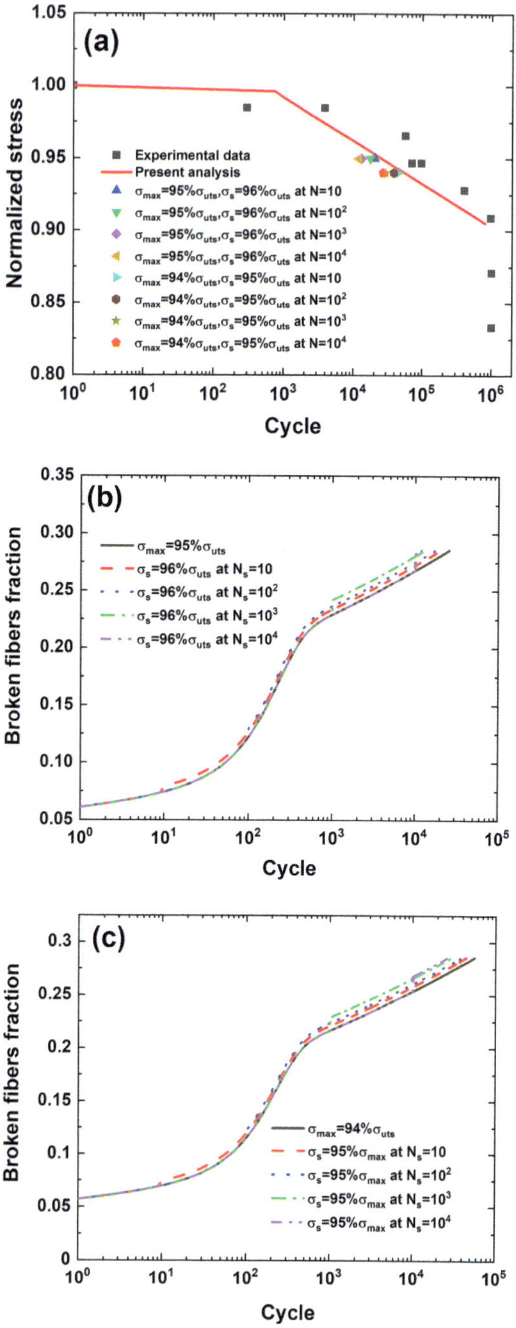

Fig. 4.7 3D braided C/SiC composite at $T = 1500$ °C for **a** the experimental and predicted fatigue life S–N curves under stochastic overloading; **b** the broken fibers fraction versus applied cycle number under $\sigma_{max} = 95\%\sigma_{uts}$ and $\sigma_s = 96\%\sigma_{uts}$ at $N_s = 10$, 10^2, 10^3, and 10^4; and **c** the broken fibers fraction versus applied cycle number under $\sigma_{max} = 94\%\sigma_{uts}$ and $\sigma_s = 95\%\sigma_{uts}$ at $N_s = 10$, 10^2, 10^3, and 10^4

Under fatigue peak stress $\sigma_{max} = 95\%\sigma_{uts}$ and stochastic overloading stress $\sigma_s = 96\%\sigma_{uts}$ at different occurrence applied cycle number $N_s = 10$, 10^2, 10^3, and 10^4, the fatigue life decreases from $N_f = 25{,}608$ to 20,188, 17,481, 13,067, and 11,884, and the fatigue life decreasing rate is $\Lambda = 21.1$, 31.7, 48.9, and 53.6%. Under fatigue peak stress $\sigma_{max} = 94\%\sigma_{uts}$ and stochastic overloading stress $\sigma_s = 95\%\sigma_{uts}$ at different occurrence applied cycle number $N_s = 10$, 10^2, 10^3, and 10^4, the fatigue life decreases from $N_f = 55{,}487$ to 44,362, 38,696, 29,253, and 26,665, and the fatigue life decreasing rate is $\Lambda = 20$, 30.2, 47.2, and 51.9%.

Figures 4.7b, c and Table 4.7 show the predicted broken fibers fraction under different stochastic overloading stress and occurrence applied cycles. The broken fibers fraction increases when stochastic overloading stress occurs. The difference of broken fibers fraction between original fatigue peak stress and stochastic overloading stress increases with occurrence applied cycle number.

Under fatigue peak stress $\sigma_{max} = 95\%\sigma_{uts}$ and stochastic overloading stress $\sigma_s = 96\%\sigma_{uts}$ at different occurrence applied cycle number $N_s = 10$, 10^2, 10^3, and 10^4, the broken fibers fraction increases from $P_f = 7.4\%$ to 7.9% at $N_s = 10$, from $P_f = 12.2\%$ to 12.9% at $N_s = 10^2$, from $P_f = 22.8\%$ to 24.1% at $N_s = 10^3$, and from $P_f = 26.7\%$ to 28.2% at $N_s = 10^4$. Under fatigue peak stress $\sigma_{max} = 94\%\sigma_{uts}$

Table 4.7 Fatigue life and broken fibers fraction of 3D braided C/SiC composite under stochastic loading at $T = 1500$ °C

$\sigma_{max}/\sigma_{uts} = 95\%$	Cycle number	1	10	10^2	10^3	10^4	25,608
	$P_f/(\%)$	6.09	7.43	12.2	22.9	26.8	28.6
$\sigma_s/\sigma_{uts} = 96\%$ $N_s = 10$	Cycle number	1	10	10^2	10^3	10^4	20,188
	$P_f/(\%)$	6.09	7.89	12.7	23.3	27.3	28.6
$\sigma_s/\sigma_{uts} = 96\%$ $N_s = 10^2$	Cycle number	1	10	10^2	10^3	10^4	17,481
	$P_f/(\%)$	6.09	7.43	12.9	23.6	27.5	28.6
$\sigma_s/\sigma_{uts} = 96\%$ $N_s = 10^3$	Cycle number	1	10	10^2	10^3	10^4	13,067
	$P_f/(\%)$	6.09	7.43	12.2	24.2	28.1	28.6
$\sigma_s/\sigma_{uts} = 96\%$ $N_s = 10^4$	Cycle number	1	10	10^2	10^3	10^4	11,884
	$P_f/(\%)$	6.09	7.43	12.2	22.9	28.2	28.6
$\sigma_{max}/\sigma_{uts} = 94\%$	Cycle number	1	10	10^2	10^3	10^4	55,487
	$P_f/(\%)$	5.73	6.99	11.51	21.6	25.4	28.6
$\sigma_s/\sigma_{uts} = 95\%$ $N_s = 10$	Cycle number	1	10	10^2	10^3	10^4	44,362
	$P_f/(\%)$	5.73	7.43	11.95	22.1	25.8	28.6
$\sigma_s/\sigma_{uts} = 95\%$ $N_s = 10^2$	Cycle number	1	10	10^2	10^3	10^4	38,696
	$P_f/(\%)$	5.73	6.99	12.2	22.3	26.1	28.6
$\sigma_s/\sigma_{uts} = 95\%$ $N_s = 10^3$	Cycle number	1	10	10^2	10^3	10^4	29,253
	$P_f/(\%)$	5.73	6.99	11.5	22.9	26.6	28.6
$\sigma_s/\sigma_{uts} = 95\%$ $N_s = 10^4$	Cycle number	1	10	10^2	10^3	10^4	26,665
	$P_f/(\%)$	5.73	6.99	11.5	21.6	26.8	28.6

and stochastic overloading stress $\sigma_s = 95\%\sigma_{uts}$ at different occurrence applied cycle number $N_s = 10$, 10^2, 10^3, and 10^4, the broken fibers fraction increases from $P_f =$ 6.9% to 7.4% at $N_s = 10$, from $P_f = 11.5\%$ to 12.2% at $N_s = 10^2$, from $P_f = 21.6\%$ to 22.8% at $N_s = 10^3$, and from $P_f = 25.3\%$ to 26.7% at $N_s = 10^4$.

4.4 Discussions

In this section, the fatigue life decreasing rate versus occurrence applied cycle number of unidirectional, cross-ply, 2D woven, 2.5D woven, and 3D braided C/SiC composites for different stochastic overloading stresses at elevated temperature is analyzed, as shown in Fig. 4.8.

For unidirectional HT-C/SiC composite at $T = 800$ °C in air atmosphere, at the same occurrence applied cycle number, the fatigue life decreasing rate increases with stochastic overloading stress level, i.e., $\Lambda = 8.8\%$ at $\sigma_s = 92\%\sigma_{uts}$ and $N_s = 10$ versus $\Lambda = 0.47\%$ at $\sigma_s = 32\%\sigma_{uts}$ and $N_s = 10$. At the same stochastic overloading stress level, the fatigue life decreasing rate increases with occurrence applied cycle, i.e., $\Lambda = 0.47\%$ at $\sigma_s = 32\%\sigma_{uts}$ and $N_s = 10$ versus $\Lambda = 4.6\%$ at $\sigma_s = 32\%\sigma_{uts}$ and $N_s = 10^4$, as shown in Fig. 4.8a.

For cross-ply HT-C/SiC composite at $T = 800$ °C in air atmosphere, at the same occurrence applied cycle number, the fatigue life decreasing rate increases with stochastic overloading stress level, i.e., $\Lambda = 2.8\%$ at $\sigma_s = 62\%\sigma_{uts}$ and $N_s = 10$ versus $\Lambda = 0.24\%$ at $\sigma_s = 32\%\sigma_{uts}$ and $N_s = 10$. At the same stochastic overloading stress level, the fatigue life decreasing rate increases with occurrence applied cycle, i.e., $\Lambda = 0.24\%$ at $\sigma_s = 32\%\sigma_{uts}$ and $N_s = 10$ versus $\Lambda = 13.9\%$ at $\sigma_s = 32\%\sigma_{uts}$ and $N_s = 10^5$, as shown in Fig. 4.8b.

For 2D woven CVI-C/SiC composite at $T = 550$ °C in air atmosphere, at the same occurrence applied cycle number, the fatigue life decreasing rate increases with stochastic overloading stress level, i.e., $\Lambda = 1.8\%$ at $\sigma_s = 72\%\sigma_{uts}$ and $N_s = 10$ versus $\Lambda = 0.03\%$ at $\sigma_s = 42\%\sigma_{uts}$ and $N_s = 10$. At the same stochastic overloading stress level, the fatigue life decreasing rate increases with occurrence applied cycle, i.e., $\Lambda = 0.03\%$ at $\sigma_s = 42\%\sigma_{uts}$ and $N_s = 10$ versus $\Lambda = 0.1\%$ at $\sigma_s = 42\%\sigma_{uts}$ and $N_s = 10^4$, as shown in Fig. 4.8c.

For 2.5D woven CVI-C/SiC composite at $T = 800$ °C in air atmosphere, at the same occurrence applied cycle number, the fatigue life decreasing rate increases with stochastic overloading stress level, i.e., $\Lambda = 2.3\%$ at $\sigma_s = 82\%\sigma_{uts}$ and $N_s = 10$ versus $\Lambda = 1.3\%$ at $\sigma_s = 40\%\sigma_{uts}$ and $N_s = 10$. At the same stochastic overloading stress level, the fatigue life decreasing rate increases with occurrence applied cycle, i.e., $\Lambda = 1.3\%$ at $\sigma_s = 40\%\sigma_{uts}$ and $N_s = 10$ versus $\Lambda = 86.4\%$ at $\sigma_s = 40\%\sigma_{uts}$ and $N_s = 10^5$, as shown in Fig. 4.8d.

For 3D braided CVI-C/SiC composite at $T = 1300$ °C in vacuum atmosphere, at the same occurrence applied cycle number, the fatigue life decreasing rate increases with stochastic overloading stress level, i.e., $\Lambda = 22.8\%$ at $\sigma_s = 96\%\sigma_{uts}$ and $N_s = 10$ versus $\Lambda = 21.6\%$ at $\sigma_s = 95\%\sigma_{uts}$ and $N_s = 10$. At the same stochastic

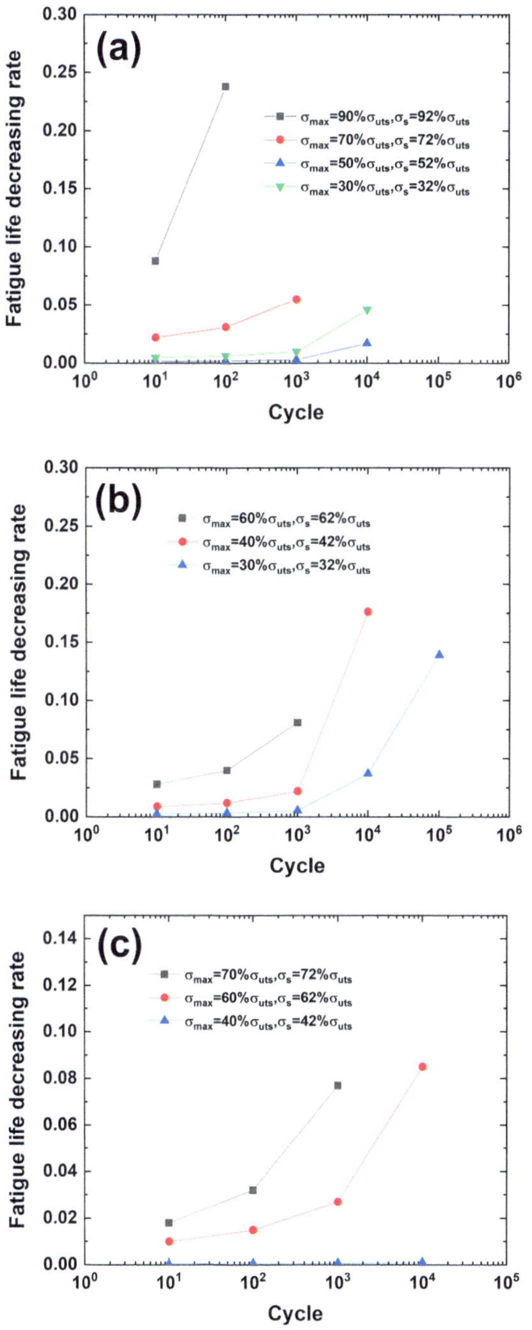

Fig. 4.8 Fatigue life decreasing rate versus occurrence applied cycle number of **a** unidirectional C/SiC at $T = 800$ °C in air atmosphere; **b** cross-ply C/SiC at $T = 800$ °C in air atmosphere; **c** 2D woven C/SiC at $T = 550$ °C in air atmosphere; **d** 2.5D woven C/SiC at $T = 800$ °C in air atmosphere; **e** 3D braided C/SiC at $T = 1300$ °C in vacuum; and **f** 3D braided C/SiC at $T = 1500$ °C in vacuum atmosphere

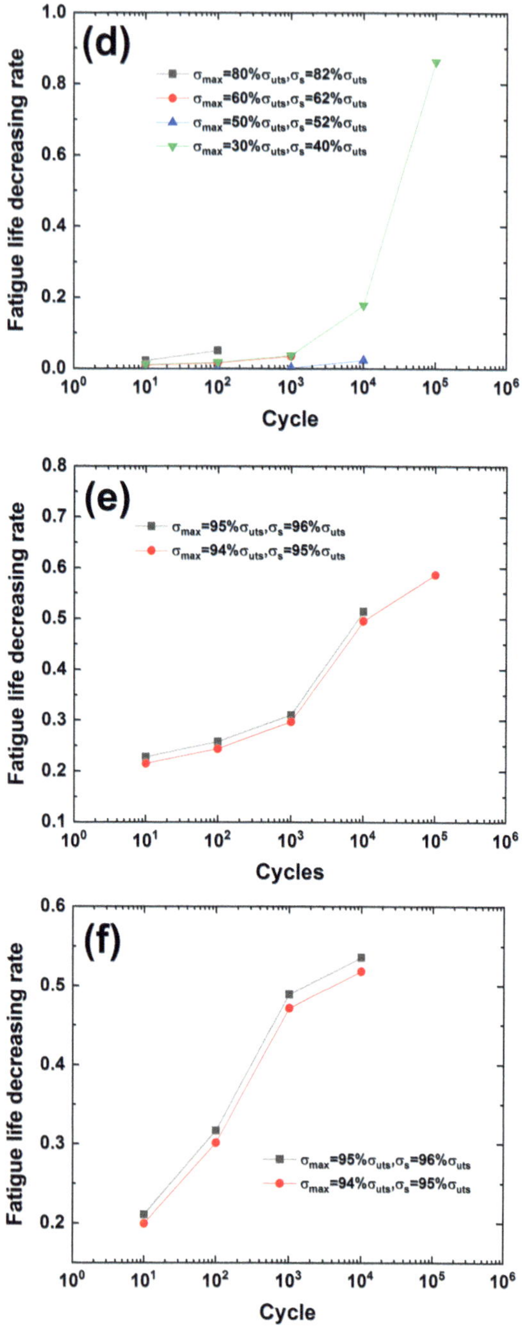

Fig. 4.8 (continued)

overloading stress level, the fatigue life decreasing rate increases with occurrence applied cycle, i.e., $\Lambda = 21.6\%$ at $\sigma_s = 95\%\sigma_{uts}$ and $N_s = 10$ versus $\Lambda = 58.7\%$ at $\sigma_s = 95\%\sigma_{uts}$ and $N_s = 10^5$, as shown in Fig. 4.8e.

For 3D braided C/SiC composite at $T = 1500$ °C in vacuum atmosphere, at the same occurrence applied cycle number, the fatigue life decreasing rate increases with stochastic overloading stress level, i.e., $\Lambda = 21.1\%$ at $\sigma_s = 96\%\sigma_{uts}$ and $N_s = 10$ versus $\Lambda = 20\%$ at $\sigma_s = 95\%\sigma_{uts}$ and $N_s = 10$. At the same stochastic overloading stress level, the fatigue life decreasing rate increases with occurrence applied cycle, i.e., $\Lambda = 20\%$ at $\sigma_s = 95\%\sigma_{uts}$ and $N_s = 10$ versus $\Lambda = 51.9\%$ at $\sigma_s = 95\%\sigma_{uts}$ and $N_s = 10^4$, as shown in Fig. 4.8f.

4.5 Conclusions

In this paper, an elevated temperature micromechanical fatigue life prediction method of fiber-reinforced CMCs subjected to stochastic overloading was developed considering multiple fatigue damage mechanisms of interface wear and interface and fibers oxidation. Relationships between stochastic overloading stress and occurrence applied cycles, broken fibers fraction, and fatigue life were established. Experimental fatigue life S–N curves of C/SiC composites with different fiber preforms were predicted for different stochastic overloading stress and occurrence cycle numbers.

1. The broken fibers fraction increased when stochastic overloading stress occurred, leading to the decrease of fatigue life.
2. Under the same stochastic overloading stress, the difference of broken fibers fraction between original fatigue peak stress and stochastic overloading stress increased with occurrence applied cycle, and the fatigue life of C/SiC decreased with increasing occurrence applied cycle.
3. Under the same occurrence applied cycle, the fatigue life of C/SiC composite decreased with increasing stochastic overloading stress.

References

Budiansky B, Hutchinson JW, Evans AG (1986) Matrix fracture in fiber-reinforced ceramics. J Mech Phys Solids 34:167–189. https://doi.org/10.1016/0022-5096(86)90035-9

Curtin WA (1991) Theory of mechanical properties of ceramic-matrix composites. J Am Ceram Soc 74:2837–2845. https://doi.org/10.1111/j.1151-2916.1991.tb06852.x

Du SM, Qiao SR (2011) Tension-tension fatigue behavior of 3D-C/SiC composite at 1500 °C. J Mater Eng 5:34–36 (in Chinese)

Du SM, Qiao SR, Ji GC, Han D (2002) Tension-tension fatigue behavior of 3D-C/SiC composite at room temperature and 1300 °C. Mater Eng 9:22–25 (in Chinese)

Evans AG, Zok FW, McMeeking RM (1995) Fatigue of ceramic matrix composites. Acta Metal Mater 43:859–875. https://doi.org/10.1016/0956-7151(94)00304-Z

Lara-Curzio E (1999) Analysis of oxidation-assisted stress-rupture of continuous fiber-reinforced ceramic matrix composites at intermediate temperatures. Compos Part A 30:549–554. https://doi.org/10.1016/S1359-835X(98)00148-1

Lee SS, Stinchcomb WW (1994) Damage mechanisms of cross-ply Nicalon/CAS-II laminate under cyclic tension. Ceram Eng Sci Proc 15:40–48. https://doi.org/10.1002/9780470314500.ch5

Li LB (2015a) Tension-tension fatigue behavior of unidirectional C/SiC ceramic-matrix composite at room and elevated temperature and 800 °C in air atmosphere. Mater 8:3316–3333. https://doi.org/10.3390/ma8063316

Li LB (2015b) Damage evolution and life prediction of cross-ply C/SiC ceramic-matrix composite under cyclic fatigue loading at room temperature and 800°C in air. Mater 8:8539–8560. https://doi.org/10.3390/ma8125474

Li LB (2015c) Fatigue life prediction of carbon fiber-reinforced ceramic-matrix composites at room and elevated temperatures. Part II: Experimental comparisons. Appl Compos Mater 22:961–972. https://doi.org/10.1007/s10443-015-9445-4

Li LB (2016a) Fatigue life prediction of fiber-reinforced ceramic-matrix composites with different fiber preforms at room and elevated temperatures. Mater 9:207. https://doi.org/10.3390/ma9030207

Li LB (2016b) Fatigue life prediction of carbon fiber-reinforced ceramic-matrix composites at room and elevated temperatures. Part I: Experimental analysis. Appl Compos Mater 23:101–117. https://doi.org/10.1007/s10443-015-9446-3

Li LB (2019a) Thermomechanical fatigue of ceramic-matrix composites. Wiley-VCH, Weinheim, Germany

Li LB (2019b) Stress-rupture of fiber-reinforced ceramic-matrix composites with stochastic loading at intermediate temperatures Part I: theoretical analysis. Mater 12:3123. https://doi.org/10.3390/ma12193123

Li LB (2020) Durability of ceramic matrix composites. Woodhead Publishing, Oxford, UK. 2020. https://doi.org/10.1016/C2018-0-03346-0

Liu H, Yuan H, Yang Z (2020) Damage modeling of oxide/oxide ceramic matrix composites under cyclic loading conditions. Ceram Int 46:23379–23389. https://doi.org/10.1016/j.ceramint.2020.05.208

Li LB, Reynaud P, Fantozzi G (2020) Cyclic-dependent damage evolution in self-healing woven SiC/[Si-B-C] ceramic-matrix composites at elevated temperatures. Mater 13:1478. https://doi.org/10.3390/ma13061478

Mall S, Engesser JM (2006) Effects of frequency on fatigue behavior of CVI C/SiC at elevated temperature. Compos Sci Technol 66:863–874. https://doi.org/10.1016/j.compscitech.2005.06.020

Min JB, Xue D, Shi Y (2014) Micromechanics modeling for fatigue damage analysis designed for fabric reinforced ceramic matrix composites. Compos Struct 111:213–223. https://doi.org/10.1016/j.compstruct.2013.12.025

Mizuno M, Zhu SJ, Nagano Y, Sakaida Y, Kagawa Y, Watanabe M (1996) Cyclic-fatigue behavior of SiC/SiC composites at room and high temperatures. J Am Ceram Soc 79:3065–3018. https://doi.org/10.1111/j.1151-2916.1996.tb08078.x

Naslain R (2004) Design, preparation and properties of non-oxide CMCs for application in engines and nuclear reactors: an overview. Compos Sci Technol 64:155–170. https://doi.org/10.1016/S0266-3538(03)00230-6

Ohnabe H, Masaki S, Onozuka M, Miyahara K, Sasa T (1999) Potential application of ceramic matrix composites to aero-engine components. Compos Part A 30:489–496. https://doi.org/10.1016/S1359-835X(98)00139-0

Quiney Z, Weston E, Nocholson PI, Pattison S, Bache MR (2020) Volumetric assessment of fatigue damage in a SiC_f/SiC ceramic matrix composite via in situ X-ray computed tomography. J Euro Ceram Soc 40:3788–3794. https://doi.org/10.1016/j.jeurceramsoc.2020.04.037

Reynaud P (1996) Cyclic fatigue of ceramic-matrix composites at ambient and elevated temperatures. Compos Sci Technol 56:809–814. https://doi.org/10.1016/0266-3538(96)00025-5

Ruggles-Wrenn MB, Christensen DT, Chamberlain AL, Lane JE, Cook TS (2011) Effect of frequency and environment on fatigue behavior of a CVI SiC/SiC ceramic matrix composite at 1200 °C. Compos Sci Technol 71:190–196. https://doi.org/10.1016/j.compscitech.2010.11.008

Ruggles-Wrenn MB, Delapasse J, Chamberlain AL, Lane JE, Cook TS (2012) Fatigue behavior of a Hi-Nicalon^{TM}/SiC-B_4C composite at 1200°C in air and in steam. Mater Sci Eng A 534:119–128. https://doi.org/10.1016/j.msea.2016.09.076

Suzuki T, Mahfuz H, Takanashi M (2019) A new stiffness degradation model for fatigue life prediction of GFRPs under random loading. Int J Fatigue 119:220–228. https://doi.org/10.1016/j.ijfatigue.2018.09.021

Yang F (2011) Research on fatigue behavior of 2.5d woven ceramic matrix composites. Master Thesis, Nanjing: Nanjing University of Aeronautics and Astronautics, 2011. (in Chinese)

Zhu D (2018) Aerospace ceramic materials: thermal, environmental barrier coatings and SiC/SiC ceramic matrix composites for turbine engine applications. NASA/TM-2018-219884

Chapter 5
Fatigue Damage and Fracture of Ceramic-Matrix Composites Subjected to Stochastic Loading

Abstract In this chapter, a micromechanical fatigue limit stress model of fiber-reinforced ceramic-matrix composites (CMCs) subjected to stochastic overloading stress is developed. Fatigue limit stress of different C/SiC composites (i.e., unidirectional (UD), cross-ply (CP), 2D, 2.5D, and 3D C/SiC) is predicted based on the micromechanical fatigue damage models and fatigue failure criterion. Under cyclic fatigue loading, the fatigue damage and fracture under stochastic overloading stress at a different applied cycle number are characterized using two parameters of fatigue life decreasing rate and broken fibers fraction. Relationships between fatigue life decreasing rate, stochastic overloading stress, corresponding occurrence applied cycle number, and broken fibers fraction are analyzed.

Keywords Ceramic-matrix composites (CMCs) · Fatigue · Stochastic loading · Fatigue limit stress

5.1 Introduction

Under cyclic fatigue loading, matrix cracking, interface debonding, interface wear, and fibers fracture occur with applied cycle, and these fatigue damage mechanisms degrade the mechanical performance of fiber-reinforced CMCs (Naslain 2004; Li 2015a, 2019a, 2020a, b; Reynaud 1996; Zhu et al. 1999). Fatigue limit stress is a key parameter for design of CMC components. However, the fatigue limit stress of CMCs depends on many factors, i.e., fiber characteristic and fiber properties (Yang 2011; Du et al. 2009), loading frequency (Shuler et al. 1993; Ruggles-Wrenn et al. 2011), temperature (Li et al. 2020; Li 2015b), and testing conditions (Ruggles-Wrenn and Sharma 2011; Ruggles-Wrenn and Lee 2016; Ruggles-Wrenn et al. 2018). Under cyclic fatigue loading, stochastic overloading stress may occur due to the special operation condition of aeroengine, which can affect the internal fatigue damage evolution and lifetime of CMCs (Suzuki et al. 2019; Li 2019b). Reynaud (1996), Evans (1997), and Li (2015c, 2016a, b) developed micromechanical fatigue life prediction methods for fiber-reinforced CMCs considering different fatigue damage mechanisms. The degradation rate of the fiber/matrix interface shear stress and fiber strength affects the fatigue life and fatigue limit stress. However, in the developed

L. Li, *Damage and Fracture of Ceramic-Matrix Composites Under Stochastic Loading*, Advanced Ceramics and Composites 2,
https://doi.org/10.1007/978-981-16-2141-3_5

micromechanical model, the effect of stochastic overloading stress on fatigue limit stress has not been considered.

In this chapter, a micromechanical fatigue limit stress model of fiber-reinforced CMCs subjected to stochastic overloading stress is developed. Fatigue limit stress for different C/SiC composites is predicted based on the micromechanical fatigue damage models and fatigue failure criterion. Relationships between fatigue life decreasing rate, stochastic overloading stress, corresponding occurrence applied cycle number, and broken fiber fraction are analyzed.

5.2 Theoretical Model

When stochastic overloading stress occurs under cyclic fatigue loading, the fatigue damage evolution of matrix cracking, interface debonding, and fiber failure is affected. Figure 5.1 shows the stochastic overloading stress σ_s occurred at different applied cycle number. In the present analysis, the overloading stress σ_s remains the same at applied cycle number N_1, N_2, and N_3.

Based on the global load sharing (GLS) criterion, under stochastic overloading stress, the stress carried by intact and broken fiber is determined by Eq. (5.1) (Curtin 1991)

$$\frac{\sigma_s}{V_f} = \Phi_s(1 - q) + \Phi_b q \tag{5.1}$$

where V_f is the fiber volume, Φ_s is the intact fiber stress under stochastic overloading stress, Φ_b is the stress carried by broken fiber, and q is the fiber failure probability and can be determined by Eq. (5.2).

$$q = 1 - \exp\left[-\Theta\,\Omega^{m_f}\left(\frac{\Phi_s}{\sigma_{fc}}\right)^{m_f+1}\right] \tag{5.2}$$

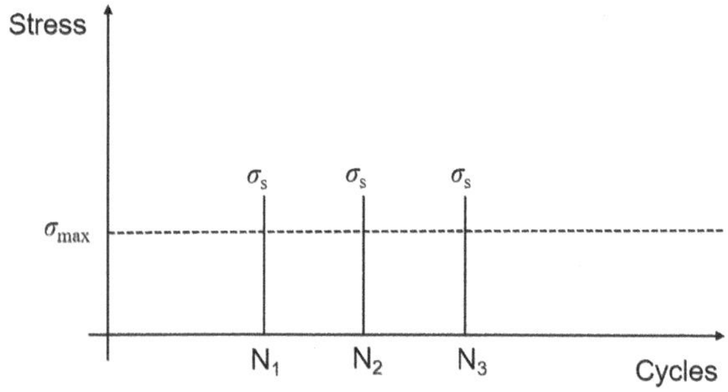

Fig. 5.1 Stochastic overloading stress σ_s occurred at different applied cycle number

$$\Phi_b = \frac{\Phi_s}{\Theta\,\Omega^{m_f} q}\left(\frac{\sigma_{fc}}{\Phi_s}\right)^{m_f+1}\left\{1-\exp\left[-\Theta\,\Omega^{m_f}\left(\frac{\Phi_s}{\sigma_{fc}}\right)^{m_f+1}\right]\right\} - \frac{\Phi_s}{q}\exp\left[-\Theta\,\Omega^{m_f}\left(\frac{\Phi_s}{\sigma_{fc}}\right)^{m_f+1}\right] \tag{5.3}$$

where m_f is the fiber Weibull modulus, σ_{fc} is the fiber characteristic strength, and Θ and Ω denote the degradation rate of interface shear stress and fiber strength and can be determined by Eqs. (5.4) and (5.5), respectively.

$$\Theta(N) = \frac{1}{1-(1-\phi)\left[1-\exp(-\delta N^{\lambda})\right]} \tag{5.4}$$

$$\Omega(N) = \frac{1}{1-p_1(\log N)^{p_2}} \tag{5.5}$$

where ϕ is the ratio between steady interface shear stress and initial interface shear stress, δ and λ are the interface wear model parameter, and p_1 and p_2 are the fiber strength degradation model parameter.

Substituting Eqs. (5.2) and (5.3) into Eq. (5.1), the relation between the applied stress and fiber intact stress is determined by Eq. (5.6).

$$\frac{\sigma_s}{V_f} = \frac{\Phi_s}{\Theta\,\Omega^{m_f}}\left(\frac{\sigma_{fc}}{\Phi_s}\right)^{m_f+1}\left\{1-\exp\left[-\Theta\,\Omega^{m_f}\left(\frac{\Phi_s}{\sigma_{fc}}\right)^{m_f+1}\right]\right\} \tag{5.6}$$

Using Eqs. (5.4)–(5.6), the intact fiber stress under stochastic overloading stress can be obtained with occurrence applied cycle number. Substituting the intact fiber stress under stochastic overloading stress into Eq. (5.2), the fraction of broken fiber under stochastic overloading stress can be obtained. Under cyclic fatigue loading, when the fiber failure probability approaches critical value, the composite fatigue fractures. The fatigue limit stress of CMCs at room temperature can be obtained using the developed life prediction model and fatigue limit cycle number N_{limit}.

The fatigue life decreasing rate is

$$\Lambda = \frac{N_f(\sigma_{limit}) - N_f(\sigma_s)}{N_f(\sigma_{limit})} \tag{5.7}$$

where $N_f(\sigma_{limit})$ is the fatigue failure cycle number under fatigue limit stress, and $N_f(\sigma_s)$ is the fatigue failure cycle number under stochastic overloading stress.

5.3 Experimental Comparisons

Under cyclic fatigue loading, the stochastic overloading stress affects fatigue damage evolution, i.e., increasing fiber failure probability, and decreasing fatigue life. In this section, the fatigue limit stress of different C/SiC composites is predicted. Material properties and fatigue damage model parameters of C/SiC composite are listed in Table 5.1. Under fatigue limit stress, the stochastic overloading occurred in different applied cycle number can decrease the fatigue life. Using the developed fiber failure model in Eq. (5.2) and the fatigue damage models in Eqs. (5.4) and (5.5), the effect of stochastic overloading stress level and corresponding occurrence cycle number on fatigue limit stress and corresponding fatigue life is analyzed. The relationships between stochastic overloading stress level, occurrence cycle number, broken fiber fraction, and fatigue limit stress are established.

5.3.1 *Unidirectional C/SiC Composite*

The unidirectional T-700™ carbon fiber-reinforced silicon carbide composite was fabricated using the hot-pressing (HP) method. Low-pressure chemical vapor infiltration was employed to deposit approximately 5–20 layer PyC/SiC with the mean

Table 5.1 Material properties of C/SiC composite

Items	Unidirectional (Li 2015a)	Cross-ply (Li 2015b)	2D (Shuler et al. 1993)	2.5D (Yang 2011)	3D (Du et al. 2009)
Manufacturing process	Hot pressing	Hot pressing	Chemical vapor infiltration (CVI)	CVI	CVI
Stress ratio	0.1	0.1	0.1	0.1	0.1
Frequency/(Hz)	10	10	10	10	60
Fiber type	T-700™	T-700™	T-300™	T-300™	T-300™
V_f	0.4	0.4	0.45	0.4	0.4
σ_{uts}/(MPa)	270	124	420	225	276
r_f/(μm)	3.5	3.5	3.5	3.5	3.5
τ_{io}/(MPa)	8	6.2	25	20	20
τ_{imin}/(MPa)	0.3	1.5	8	8	5
δ	0.04	0.06	0.002	0.001	0.02
λ	1.5	1.8	1.0	1.0	1.0
p_1	0.01	0.01	0.018	0.02	0.012
p_2	1.0	0.8	1.0	1.2	1.0
m_f	5	5	5	5	5

thickness of 0.2 μm. The nano-SiC powder and sintering additives were ball milled for 4 h using SiC balls. After drying, the powders were dispersed in xylene with polycarbonsilane (PCS) to form the slurry. Carbon fiber tows were infiltrated by the slurry and wound to form aligned unidirectional composite sheets. After drying, the sheets were cut to a size of 150 mm × 150 mm and pyrolyzed in argon. Then, the sheets were stacked in a graphite die and sintered by hot pressing. The dog-bone-shaped specimens were cut from 150 mm × 150 mm panels by water cutting. The tension–tension fatigue tests were conducted on a MTS Model 809 servo hydraulic load-frame (MTS Systems Corp., Minneapolis, MN, USA). The fatigue experiments were in a sinusoidal waveform with a loading frequency $f = 10$ Hz. The fatigue load ratio ($\sigma_{min}/\sigma_{max}$) was $R = 0.1$. The fatigue tests were conducted under load control at room temperature.

Figure 5.2 shows the experimental and predicted fatigue life S–N curves of unidirectional C/SiC composite. When the fatigue limit applied cycle number is defined to be $N_{limit} = 10^6$, the corresponding predicted fatigue limit stress is approximately $\sigma_{limit} = 241$ MPa (approximately 89.2%σ_{uts}).

Figure 5.2b shows the fatigue life decreasing rate versus the occurrence cycle number of stochastic overloading curves for different stochastic overloading stress levels $\sigma_s = 245$, 250, and 255 MPa (i.e., approximately 1.016, 1.037, 1.058 of fatigue limit stress). During the application of CMC components, the overloading stress level is not high, and the low overloading stress level is chosen for analysis. Under the same stochastic overloading stress level $\sigma_s = 245$, 250, and 255 MPa, the fatigue life decreasing rate increases with the occurrence applied cycle of stochastic overloading (i.e., $N_s = 10$, 10^2, 10^3, 10^4, and 10^5). Under $\sigma_s = 245$ MPa, the fatigue life decreasing rate increases from $\Lambda = 0.14392$ at $N_s = 10$ to $\Lambda = 0.8074$ at $N_s = 10^5$; under $\sigma_s = 250$ MPa, the fatigue life decreasing rate increases from $\Lambda = 0.3091$ at $N_s = 10$ to $\Lambda = 0.97479$ at $N_s = 10^4$; and under $\sigma_s = 255$ MPa, the fatigue life decreasing rate increases from $\Lambda = 0.4559$ at $N_s = 10$ to $\Lambda = 0.9966$ at $N_s = 10^3$. When applied cycle is between $N_s = 10$ and 10^2, the fatigue life decreasing rate increases rapidly with occurrence applied cycle; however, when applied cycle is higher than $N_s = 10^2$, the fatigue life decreasing rate increases slowly with applied cycle. At the initial stage of cyclic fatigue loading, the fatigue damage mechanisms of matrix cracking, interface debonding, and wear depend on the fatigue peak stress level. The occurrence of stochastic overloading stress at initial stage of cyclic fatigue loading deteriorates fatigue damage evolution, i.e., decreasing matrix crack spacing, increasing interface debonding length, and broken fiber fraction; however, when matrix cracking and interface wear approach steady-state, the effect of stochastic overloading stress on fatigue damage or fatigue life decreasing rate decreases.

Figure 5.2c–e shows the broken fiber faction versus applied cycle number curves for different stochastic overloading stress levels and occurrence applied cycle number. The broken fiber fraction increases when stochastic overloading stress occurs, and the difference of broken fiber fraction between original peak stress and stochastic overloading stress level increases with applied cycle number.

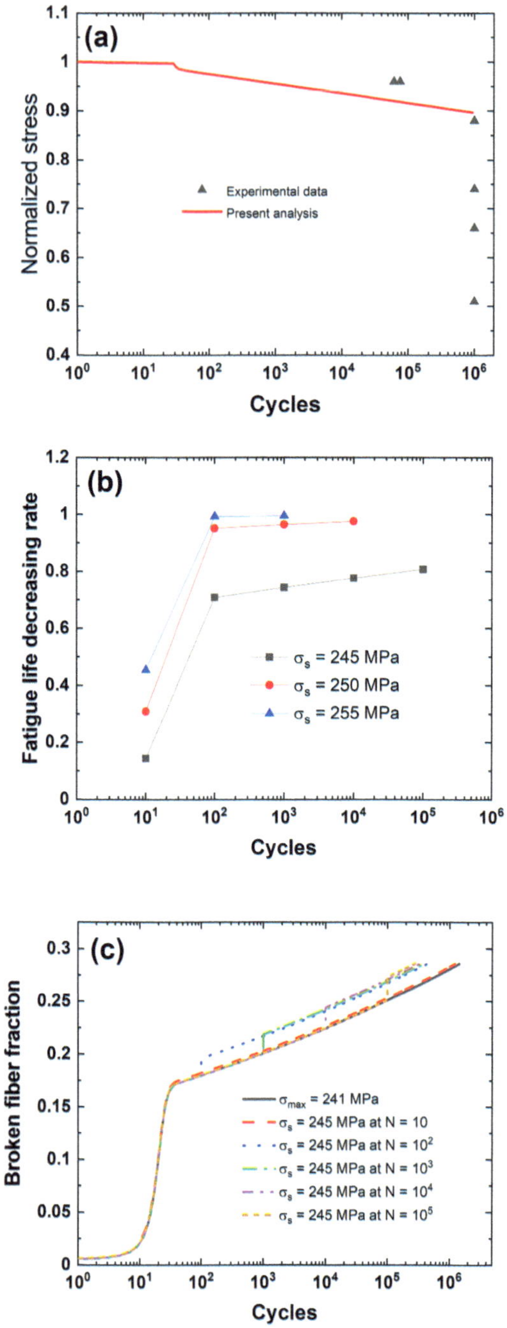

Fig. 5.2 **a** Experimental and predicted fatigue life S–N curves; **b** the fatigue life decreasing rate versus occurrence applied cycle number curve for different stochastic loading stress levels; **c** the broken fibers fraction versus applied cycle number curves under $\sigma_{\text{limit}} = 241$ MPa and stochastic overloading stress $\sigma_s = 245$ MPa at $N_s = 10$, 10^2, 10^3, 10^4, and 10^5; **d** the broken fibers fraction versus applied cycle number curves under $\sigma_{\text{limit}} = 241$ MPa and stochastic overloading stress $\sigma_s = 250$ MPa at $N_s = 10$, 10^2, 10^3, 10^4, and 10^5; and **e** the broken fibers fraction versus applied cycle number curves under $\sigma_{\text{limit}} = 241$ MPa and stochastic overloading stress $\sigma_s = 255$ MPa at $N_s = 10$, 10^2, and 10^3 of unidirectional C/SiC composite

Fig. 5.2 (continued)

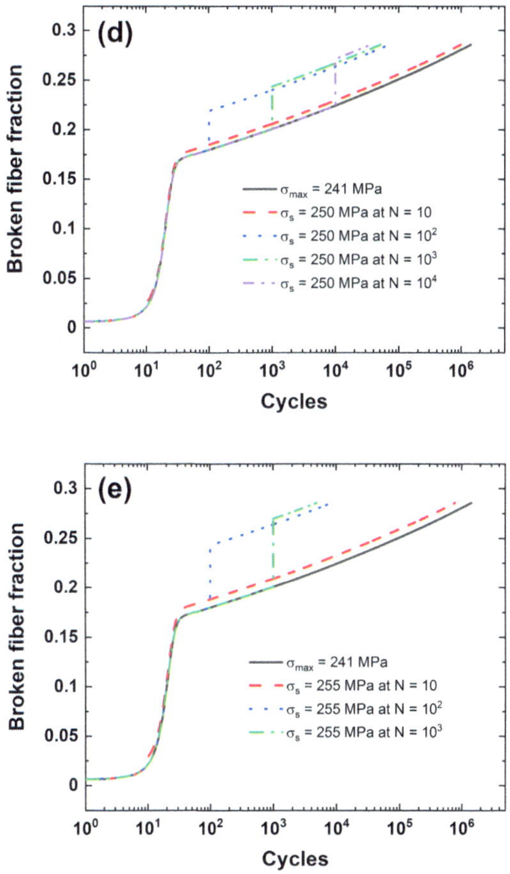

Table 5.2 shows the fatigue limit stress and broken fiber fraction at different occurrence cycle number and stochastic overloading stress. When stochastic overloading stress $\sigma_s = 245$ MPa occurs at applied cycle $N_s = 10$, 10^2, 10^3, 10^4, and 10^5, the broken fiber fraction increases from $q = 0.02113$, 0.17991, 0.20077, 0.22431, and 0.25085 under $\sigma_{\text{limit}} = 241$ MPa to $q = 0.02329$, 0.19662, 0.21915, 0.2445, and 0.27298; when stochastic overloading stress $\sigma_s = 250$ MPa occurs at applied cycle $N_s = 10$, 10^2, 10^3, and 10^4, the broken fiber fraction increases to $q = 0.02626$, 0.21897, 0.24365, and 0.27131; and, when stochastic overloading stress $\sigma_s = 255$ MPa occurs at applied cycle $N = 10$, 10^2, and 10^3, the broken fiber fraction increases to $q =$ 0.02952, 0.24295, and 0.26984.

Table 5.2 Fatigue limit stress and broken fibers fraction of unidirectional C/SiC composite under stochastic overloading stress

σ_{max} = 241 MPa	N_f		$N = 1$	$N = 10$	$N = 10^2$	$N = 10^3$	$N = 10^4$	$N = 10^5$
	1,431,993	q	0.00609	0.02113	0.17991	0.20077	0.22431	0.25085
σ_s = 245 MPa $N = 10$	N_f		$N = 1$	$N = 10$	$N = 10^2$	$N = 10^3$	$N = 10^4$	$N = 10^5$
	1,225,895	q	0.00609	0.02329	0.18207	0.20294	0.22648	0.25302
σ_s = 245 MPa $N = 10^2$	N_f		$N = 1$	$N = 10$	$N = 10^2$	$N = 10^3$	$N = 10^4$	$N = 10^5$
	418,087	q	0.00609	0.02113	0.19662	0.21749	0.24103	0.26757
σ_s = 245 MPa $N = 10^3$	N_f		$N = 1$	$N = 10$	$N = 10^2$	$N = 10^3$	$N = 10^4$	$N = 10^5$
	368,186	q	0.00609	0.02113	0.17991	0.21915	0.24269	0.26923
σ_s = 245 MPa $N = 10^4$	N_f		$N = 1$	$N = 10$	$N = 10^2$	$N = 10^3$	$N = 10^4$	$N = 10^5$
	320,486	q	0.00609	0.02113	0.17991	0.20077	0.2445	0.27104
σ_s = 245 MPa $N = 10^5$	N_f		$N = 1$	$N = 10$	$N = 10^2$	$N = 10^3$	$N = 10^4$	$N = 10^5$
	275,798	q	0.00609	0.02113	0.17991	0.20077	0.22431	0.27298
σ_s = 250 MPa $N = 10$	N_f		$N = 1$	$N = 10$	$N = 10^2$	$N = 10^3$	$N = 10^4$	$N = 10^5$
	989,365	q	0.00609	0.02626	0.18504	0.2059	0.22944	0.25598
σ_s = 250 MPa $N = 10^2$	N_f		$N = 1$	$N = 10$	$N = 10^2$	$N = 10^3$	$N = 10^4$	$N = 70{,}700$
	70,700	q	0.00609	0.02113	0.21897	0.23983	0.26338	0.28571
σ_s = 250 MPa $N = 10^3$	N_f		$N = 1$	$N = 10$	$N = 10^2$	$N = 10^3$	$N = 10^4$	$N = 51{,}311$
	51,311	q	0.00609	0.02113	0.17991	0.24365	0.26719	0.28571
σ_s = 250 MPa $N = 10^4$	N_f		$N = 1$	$N = 10$	$N = 10^2$	$N = 10^3$	$N = 10^4$	$N = 36{,}095$
	36,095	q	0.00609	0.02113	0.17991	0.20077	0.27131	0.28571
σ_s = 255 MPa $N = 10$	N_f		$N = 1$	$N = 10$	$N = 10^2$	$N = 10^3$	$N = 10^4$	$N = 10^5$
	779,144	q	0.00609	0.02952	0.1883	0.20917	0.23271	0.25925
σ_s = 255 MPa $N = 10^2$	N_f		$N = 1$	$N = 10$	$N = 10^2$	$N = 10^3$	$N = 8593$	
	8593	q	0.00609	0.02113	0.24295	0.26381	0.28571	
σ_s = 255 MPa $N = 10^3$	N_f		$N = 1$	$N = 10$	$N = 10^2$	$N = 10^3$	$N = 4870$	
	4870	q	0.00609	0.02113	0.17991	0.26984	0.28571	

5.3.2 Cross-ply C/SiC Composite

The cross-ply T-700™ carbon fiber-reinforced silicon carbide composite was fabricated using the hot-pressing (HP) method, which offered the ability to fabricate dense composites via a liquid phase sintering method at a low temperature. The fiber volume is $V_f = 0.4$, and the average tensile strength is approximately $\sigma_{uts} = 124$ MPa. The dog-bone-shaped specimens were cut from 150 mm × 150 mm panels by water cutting. The tension–tension fatigue tests were conducted on a MTS Model 809 servo hydraulic load-frame (MTS Systems Corp., Minneapolis, MN, USA). The fatigue experiments were in a sinusoidal waveform with a loading frequency $f = 10$ Hz. The fatigue load ratio ($\sigma_{min}/\sigma_{max}$) was $R = 0.1$. The fatigue tests were conducted under load control at room temperature.

Figure 5.3a shows experimental and predicted fatigue life S–N curves of cross-ply C/SiC composite. When the fatigue limit applied cycle number is defined to be $N_{limit} = 10^6$, the corresponding predicted fatigue limit stress is approximately $\sigma_{limit} = 103$ MPa (approximately $83\%\sigma_{uts}$).

Figure 5.3b shows the fatigue life decreasing rate versus occurrence applied cycle number of stochastic overloading stress curves under $\sigma_s = 105$ MPa (approximately 1.019 fatigue limit stress). During the application of CMC components, the overloading stress level is not high, and the low overloading stress level is chosen for analysis. The fatigue life decreasing rate increases with the occurrence applied cycle number of stochastic overloading. Under $\sigma_s = 105$ MPa, the fatigue life decreasing rate increases from $\Lambda = 0.97114$ at $N_s = 10$ to $\Lambda = 0.98696$ at $N_s = 10^4$. When the occurrence applied cycle number is between $N_s = 10$ and 10^2, the fatigue life decreasing rate increases rapidly; however, when the occurrence applied cycle is between $N_s = 10^2$ and 10^4, the fatigue life decreasing rate increases slowly. The occurrence of stochastic overloading stress at initial stage of cyclic fatigue loading deteriorates fatigue damage evolution, i.e., decreasing matrix crack spacing in transverse and longitudinal plies, increasing interface debonding length, and broken fiber fraction; however, when matrix cracking and interface wear approach steady-state, the effect of stochastic overloading on fatigue damage or fatigue life decreasing rate decreases.

Figure 5.3c shows the broken fiber faction versus applied cycle number curves for different occurrence cycle number (i.e., $N_s = 10$, 10^2, 10^3, and 10^4). The broken fibers fraction increases when stochastic overloading stress occurs, and the difference of broken fiber fraction between original peak stress and stochastic overloading stress level increases with applied cycle.

Table 5.3 shows the fatigue limit stress and broken fiber fraction under stochastic overloading stress $\sigma_s = 105$ MPa at different occurrence applied cycle number. When stochastic overloading stress $\sigma_s = 105$ MPa occurs at applied cycle number of $N_s = 10$, 10^2, 10^3, and 10^4, the broken fiber fraction increases from $q = 0.20047$, 0.22748, 0.24133, and 0.25508 under $\sigma_{limit} = 103$ MPa to $q = 0.22205$, 0.25148, 0.26653, and 0.28143.

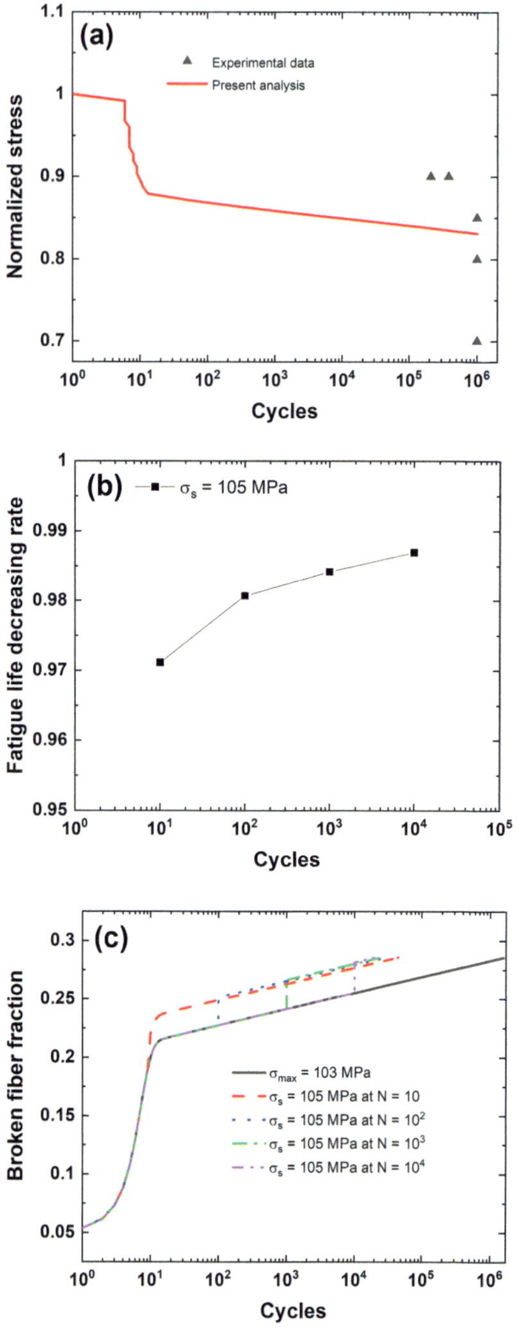

Fig. 5.3 **a** Experimental and predicted fatigue life S–N curves; **b** the fatigue life decreasing rate versus occurrence applied cycle curve; and **c** the broken fibers fraction versus applied cycle curves under $\sigma_{\text{limit}} =$ 103 MPa and stochastic overloading stress $\sigma_s =$ 105 MPa at $N_s = 10, 10^2, 10^3$, and 10^4 of cross-ply C/SiC composite

Table 5.3 Fatigue limit stress and broken fiber fraction of cross-ply C/SiC composite under stochastic overloading stress

σ_{max} =	N_f		$N = 1$	$N = 10$	$N = 10^2$	$N = 10^3$	$N = 10^4$	$N = 10^5$
103 MPa	1,568,296	q	0.05349	0.20047	0.22748	0.24133	0.25508	0.26892
σ_s =	N_f		$N = 1$	$N = 10$	$N = 10^2$	$N = 10^3$	$N = 10^4$	$N = 45{,}256$
105 MPa $N = 10$	45,256	q	0.05349	0.22205	0.24906	0.26291	0.27666	0.28571
σ_s =	N_f		$N = 1$	$N = 10$	$N = 10^2$	$N = 10^3$	$N = 10^4$	$N = 30{,}243$
105 MPa $N = 10^2$	30,243	q	0.05349	0.20047	0.25148	0.26533	0.27908	0.28571
σ_s =	N_f		$N = 1$	$N = 10$	$N = 10^2$	$N = 10^3$	$N = 10^4$	$N = 24{,}787$
105 MPa $N = 10^3$	24,787	q	0.05349	0.20047	0.22748	0.26653	0.28028	0.28571
σ_s =	N_f		$N = 1$	$N = 10$	$N = 10^2$	$N = 10^3$	$N = 10^4$	$N = 20{,}455$
105 MPa $N = 10^4$	20,455	q	0.05349	0.20047	0.22748	0.24133	0.28143	0.28571

5.3.3 2D C/SiC Composite

2D T-300™ carbon fiber-reinforced silicon carbide composite was fabricated using the chemical vapor infiltration (CVI) method. It contained 26 plies of plain-weave cloth in a [0°/90°] lay-up. Fiber preform was given a pyrolytic carbon coating. The fiber volume was 45%, and density was 1.93–1.98 g/cm^3, and the porosity was approximately 22%. The dog-bone-shaped specimens were cut from 200 mm × 200 mm panels using diamond tooling. The tension–tension fatigue tests at room temperature were conducted on a servo hydraulic load-frame that was equipped with edge-loaded grips. The fatigue experiments were performed under load control at a sinusoidal waveform and a loading frequency $f = 10$ Hz. The fatigue load ratio ($\sigma_{min}/\sigma_{max}$) was $R = 0.1$. The average tensile strength was approximately $\sigma_{uts} =$ 420 MPa.

Figure 5.4a shows the experimental and predicted fatigue life S–N curves of 2D C/SiC composite. When the fatigue limit applied cycle number is defined to be $N_{limit} = 10^6$, the corresponding predicted fatigue limit stress is approximately $\sigma_{limit} =$ 348 MPa (approximately 82.8%σ_{uts}).

Figure 5.4b shows the fatigue life decreasing rate versus the occurrence applied cycle number of stochastic overloading stress curves for different stochastic overloading stress levels of $\sigma_s = 350$, 355, and 360 MPa (i.e., approximately 1.005, 1.02, 1.034 fatigue limit stress). During the application of CMC components, the overloading stress level is not high, and the low overloading stress level is chosen for analysis. Under the same stochastic overloading stress level, the fatigue life decreasing rate increases with occurrence applied cycle of stochastic overloading. Under $\sigma_s = 350$ MPa, the fatigue life decreasing rate increases from $\Lambda = 0.03699$ at $N_s = 10$ to $\Lambda = 0.21481$ at $N_s = 10^5$; under $\sigma_s = 355$ MPa, the fatigue life decreasing rate increases from $\Lambda = 0.12826$ at $N_s = 10$ to $\Lambda = 0.59226$ at $N_s = 10^5$;

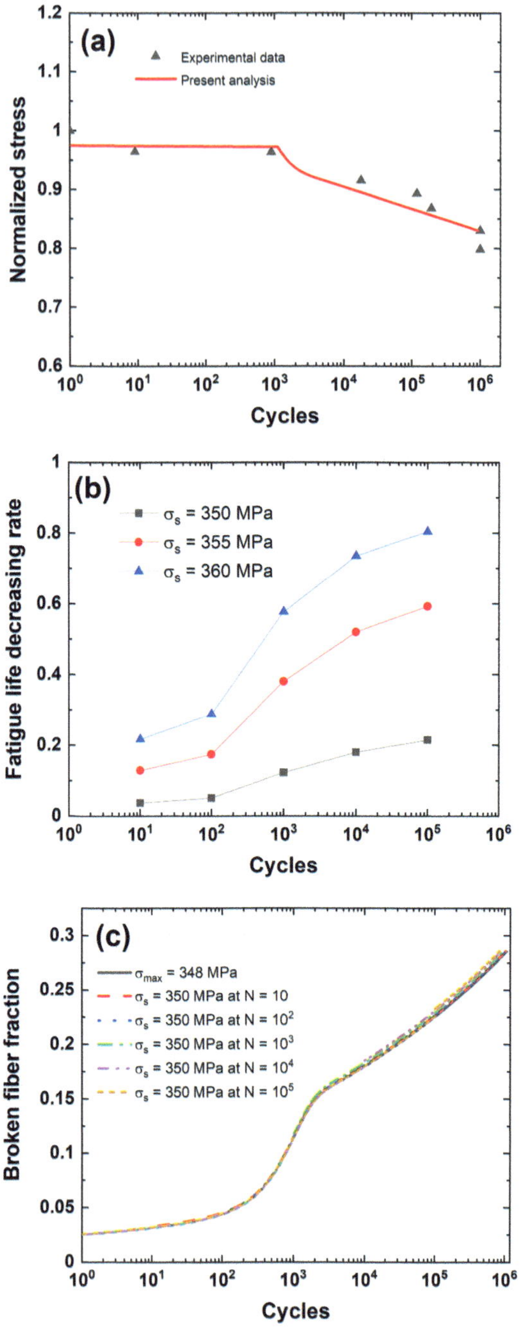

Fig. 5.4 **a** Experimental and predicted fatigue life S–N curves; **b** the fatigue life decreasing rate versus occurrence applied cycle curve for different stochastic overloading stress levels; **c** the broken fiber fraction versus applied cycle curves under $\sigma_{\text{limit}} = 348$ MPa and stochastic overloading stress $\sigma_s = 350$ MPa at $N_s = 10$, 10^2, 10^3, 10^4, and 10^5; **d** the broken fiber fraction versus applied cycle curves under $\sigma_{\text{limit}} = 348$ MPa and stochastic overloading stress $\sigma_s = 355$ MPa at $N_s = 10$, 10^2, 10^3, 10^4, and 10^5; and **e** the broken fiber fraction versus applied cycle curves under $\sigma_{\text{limit}} = 348$ MPa and stochastic overloading stress $\sigma_s = 360$ MPa at $N_s = 10$, 10^2, 10^3, 10^4, and 10^5 of 2D C/SiC composite

Fig. 5.4 (continued)

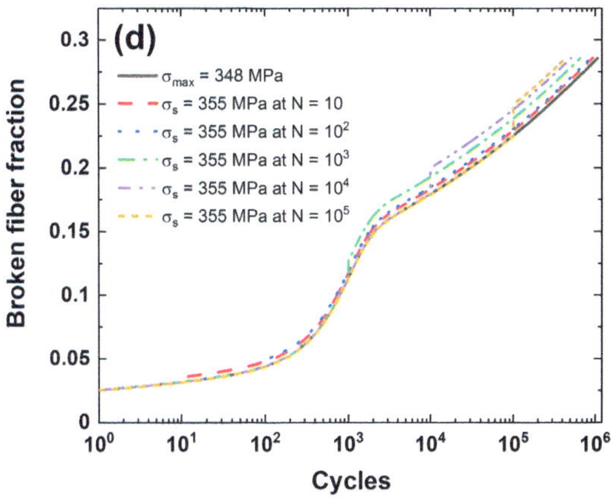

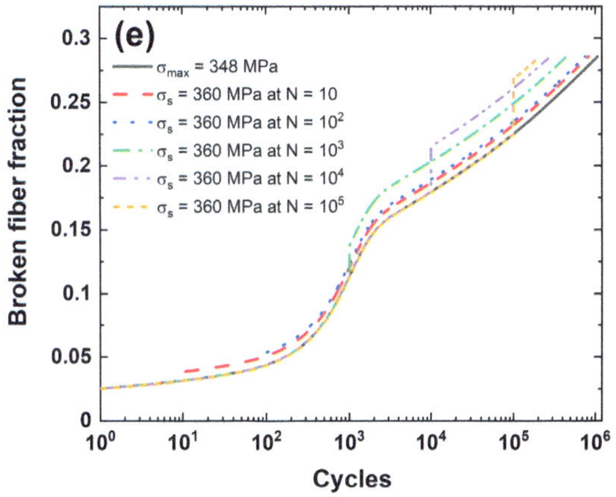

under $\sigma_s = 360$ MPa, the fatigue life decreasing rate increases from $\Lambda = 0.21723$ at $N_s = 10$ to $\Lambda = 0.80517$ at $N_s = 10^5$. When the occurrence applied cycle is between $N_s = 10$ and 10^2, the fatigue life decreasing rate increases slowly with occurrence applied cycle; however, when occurrence applied cycle is between $N_s = 10^2$ and 10^5, the fatigue life decreasing rate increases rapidly with applied cycle. For 2D C/SiC, the fatigue damage is not sensitive to stochastic overloading stress at initial stage of cyclic fatigue loading; however, with applied cycles increasing, the fatigue damage extent increases, leading to rapid increase of fatigue life decreasing rate with occurrence cycle number of stochastic overloading stress.

Figure 5.4c–e shows the broken fibers faction versus applied cycle number curves for different stochastic overloading stress levels and occurrence cycle number. The broken fiber fraction increases when stochastic overloading stress occurs, and the difference of broken fiber fraction between original peak stress and stochastic overloading stress level increases with applied cycle.

Table 5.4 shows the fatigue limit stress and broken fiber fraction at different occur-

Table 5.4 Fatigue limit stress and broken fiber fraction of 2D C/SiC composite under stochastic overloading stress

$\sigma_{\max} =$ 348 MPa	N_f		$N = 1$	$N = 10$	$N = 10^2$	$N = 10^3$	$N = 10^4$	$N = 10^5$
	1,067,612	q	0.02529	0.03153	0.04397	0.11358	0.1795	0.22527
$\sigma_s =$ 350 MPa $N = 10$	N_f		$N = 1$	$N = 10$	$N = 10^2$	$N = 10^3$	$N = 10^4$	$N = 10^5$
	1,028,121	q	0.02529	0.03261	0.04505	0.11466	0.18059	0.22635
$\sigma_s =$ 350 MPa $N = 10^2$	N_f		$N = 1$	$N = 10$	$N = 10^2$	$N = 10^3$	$N = 10^4$	$N = 10^5$
	1,013,262	q	0.02529	0.03153	0.04547	0.11508	0.18101	0.22677
$\sigma_s =$ 350 MPa $N = 10^3$	N_f		$N = 1$	$N = 10$	$N = 10^2$	$N = 10^3$	$N = 10^4$	$N = 10^5$
	937,226	q	0.02529	0.03153	0.04397	0.11731	0.18323	0.229
$\sigma_s =$ 350 MPa $N = 10^4$	N_f		$N = 1$	$N = 10$	$N = 10^2$	$N = 10^3$	$N = 10^4$	$N = 10^5$
	875,603	q	0.02529	0.03153	0.04397	0.11358	0.18516	0.23093
$\sigma_s =$ 350 MPa $N = 10^5$	N_f		$N = 1$	$N = 10$	$N = 10^2$	$N = 10^3$	$N = 10^4$	$N = 10^5$
	838,274	q	0.02529	0.03153	0.04397	0.11358	0.1795	0.23215
$\sigma_s =$ 355 MPa $N = 10$	N_f		$N = 1$	$N = 10$	$N = 10^2$	$N = 10^3$	$N = 10^4$	$N = 10^5$
	930,683	q	0.02529	0.03546	0.0479	0.11751	0.18342	0.2292
$\sigma_s =$ 355 MPa $N = 10^2$	N_f		$N = 1$	$N = 10$	$N = 10^2$	$N = 10^3$	$N = 10^4$	$N = 10^5$
	882,390	q	0.02529	0.03153	0.04941	0.11902	0.18493	0.23071
$\sigma_s =$ 355 MPa $N = 10^3$	N_f		$N = 1$	$N = 10$	$N = 10^2$	$N = 10^3$	$N = 10^4$	$N = 10^5$
	661,586	q	0.02529	0.03153	0.04397	0.12704	0.19295	0.23873
$\sigma_s =$ 355 MPa $N = 10^4$	N_f		$N = 1$	$N = 10$	$N = 10^2$	$N = 10^3$	$N = 10^4$	$N = 10^5$
	513,025	q	0.02529	0.03153	0.04397	0.11358	0.19983	0.24561
$\sigma_s =$ 355 MPa $N = 10^5$	N_f		$N = 1$	$N = 10$	$N = 10^2$	$N = 10^3$	$N = 10^4$	$N = 10^5$
	435,308	q	0.02529	0.03153	0.04397	0.11358	0.17949	0.24996
$\sigma_s =$ 360 MPa $N = 10$	N_f		$N = 1$	$N = 10$	$N = 10^2$	$N = 10^3$	$N = 10^4$	$N = 10^5$
	835,692	q	0.02529	0.0385	0.05094	0.12055	0.18646	0.23224

(continued)

rence cycle number and stochastic overloading stress. When stochastic overloading stress $\sigma_s = 350$ MPa occurs at applied cycle $N_s = 10$, 10^2, 10^3, 10^4, and 10^5, the broken fiber fraction increases from $q = 0.03153$, 0.04397, 0.11358, 0.1795, and 0.22527 under $\sigma_{\text{limit}} = 348$ MPa to $q = 0.03261$, 0.04547, 0.11731, 0.18516, and 0.23215; when stochastic overloading stress $\sigma_s = 355$ MPa occurs at applied cycle $N_s = 10$, 10^2, 10^3, 10^4, and 10^5, the broken fiber fraction increases to $q = 0.03546$,

Table 5.4 (continued)

$\sigma_s =$ 360 MPa $N = 10^2$	N_f		$N = 1$	$N = 10$	$N = 10^2$	$N = 10^3$	$N = 10^4$	$N = 10^5$
	759,497	q	0.02529	0.03153	0.05361	0.12323	0.18914	0.23491
$\sigma_s =$ 360 MPa $N = 10^3$	N_f		$N = 1$	$N = 10$	$N = 10^2$	$N = 10^3$	$N = 10^4$	$N = 10^5$
	450,695	q	0.02529	0.03153	0.04397	0.13736	0.20327	0.24905
$\sigma_s =$ 360 MPa $N = 10^4$	N_f		$N = 1$	$N = 10$	$N = 10^2$	$N = 10^3$	$N = 10^4$	$N = 10^5$
	282,293	q	0.02529	0.03153	0.04397	0.11358	0.2153	0.26108
$\sigma_s =$ 360 MPa $N = 10^5$	N_f		$N = 1$	$N = 10$	$N = 10^2$	$N = 10^3$	$N = 10^4$	$N = 10^5$
	208,007	q	0.02529	0.03153	0.04397	0.11358	0.17949	0.26861

0.04941, 0.12704, 0.19983, and 0.24996; and, when stochastic overloading stress σ_s = 360 MPa occurs at applied cycle $N = 10$, 10^2, 10^3, 10^4, and 10^5, the broken fiber fraction increases to $q = 0.0385$, 0.05361, 0.13736, 0.2153, and 0.26861.

5.3.4 2.5D C/SiC Composite

2.5D T-300™ carbon fiber-reinforced silicon carbide composite was fabricated using the chemical vapor infiltration (CVI) method. Low-pressure CVI was employed to deposit a pyrolytic carbon layer and a silicon matrix. A thin pyrolytic carbon layer was deposited on the surface of the carbon fiber as the interfacial layer with C_3H_8 at 800 °C. Methyltrichlorosilane (MTS, CH_3 $SiCl_3$) was used as gas source for the deposition of the SiC matrix. The conditions for deposition were 1000 °C. Argon was employed as a diluent gas to slow down the chemical reaction rate of deposition. The test specimens were machined from fabricated composites and further coated with SiC by isothermal CVI under same conditions. The fiber volume was $V_f = 0.4$, and the average tensile strength was approximately $\sigma_{uts} = 225$ MPa. The dog-bone-shaped specimens were cut from composite panels using diamond tooling. The tension–tension fatigue tests were conducted on a MTS Model 809 servo hydraulic load-frame (MTS Systems Corp.). The fatigue experiments were performed under load control at a loading frequency $f = 10$ Hz. The fatigue load ratio ($\sigma_{min}/\sigma_{max}$) was $R = 0.1$.

Figure 5.5a shows the experimental and predicted fatigue life S–N curves of 2.5D C/SiC composite. When the fatigue limit applied cycle number is defined to be $N_{limit} = 10^6$, the corresponding predicted fatigue limit stress is approximately $\sigma_{limit} = 143$ MPa (approximately 63.5%σ_{uts}). The fatigue limit stress of 2.5D C/SiC composite is lower than other CMCs, i.e., unidirectional, cross-ply, 2D, and 3D CMCs. The low fatigue limit stress of 2.5D C/SiC composite is mainly due to yarns bending inside of composites.

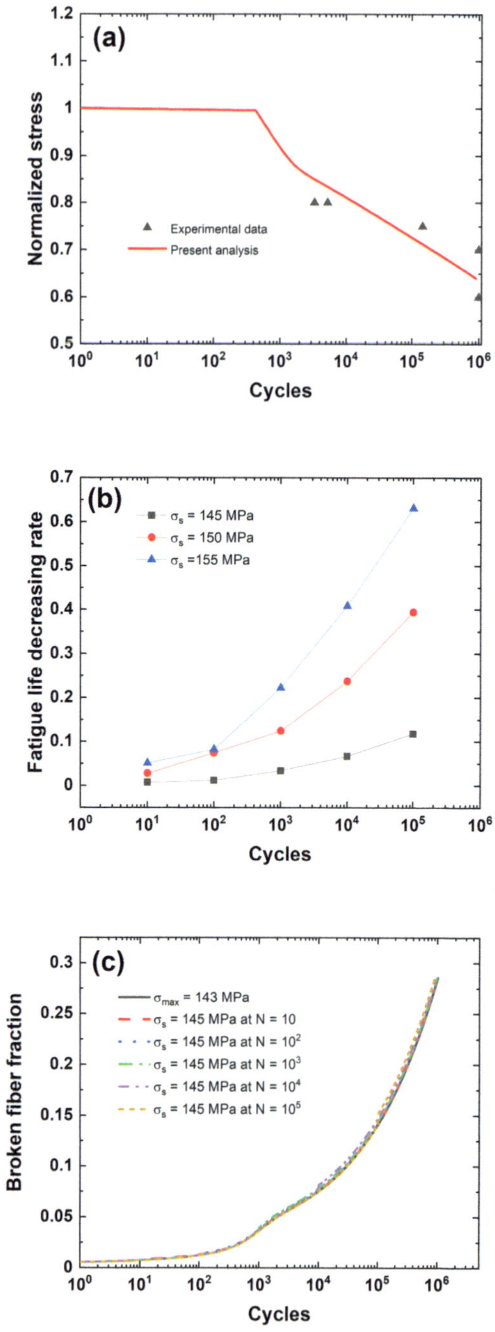

Fig. 5.5 **a** Experimental and predicted fatigue life S–N curves; **b** the fatigue life decreasing rate versus occurrence applied cycle curve for different stochastic overloading stress levels; **c** the broken fiber fraction versus applied cycle curves under $\sigma_{\text{limit}} = 143$ MPa and stochastic overloading stress of $\sigma_s = 145$ MPa at $N_s = 10, 10^2, 10^3, 10^4$, and 10^5; **d** the broken fibers fraction versus applied cycle curves under $\sigma_{\text{limit}} = 143$ MPa and stochastic overloading stress $\sigma_s = 150$ MPa at $N_s = 10, 10^2, 10^3, 10^4$, and 10^5; and **e** the broken fibers fraction versus applied cycle curves under $\sigma_{\text{limit}} = 143$ MPa and stochastic overloading stress of $\sigma_s = 155$ MPa at $N_s = 10, 10^2, 10^3, 10^4$, and 10^5 of 2.5D C/SiC composite

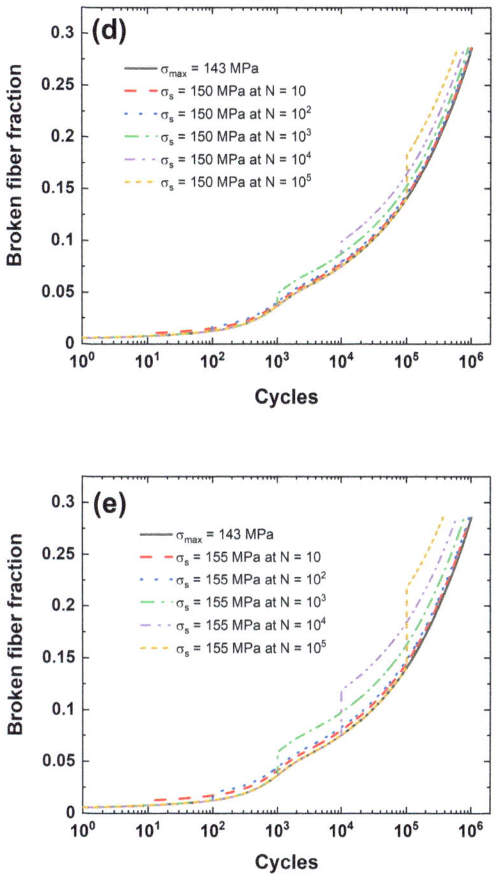

Fig. 5.5 (continued)

Figure 5.5b shows the fatigue life decreasing rate versus the occurrence applied cycle number of stochastic overloading curves for different stochastic overloading stress levels of $\sigma_s = 145$, 150, and 155 MPa (i.e., 1.014, 1.049, 1.084 fatigue limit stress). During the application of CMC components, the overloading stress level is not high, and the low overloading stress level is chosen for analysis. Under the same stochastic overloading stress level, the fatigue life decreasing rate increases with the occurrence applied cycle of stochastic overloading stress. Under $\sigma_s = 145$ MPa, the fatigue life decreasing rate increases from $\Lambda = 0.00723$ at $N_s = 10$ to $\Lambda = 0.11768$ at $N_s = 10^5$; under $\sigma_s = 150$ MPa, the fatigue life decreasing rate increases from $\Lambda = 0.02742$ at $N_s = 10$ to $\Lambda = 0.3939$ at $N_s = 10^5$; and, under $\sigma_s = 155$ MPa, the fatigue life decreasing rate increases from $\Lambda = 0.05087$ at $N_s = 10$ to $\Lambda = 0.63095$ at $N_s = 10^5$. For 2.5D C/SiC, with fatigue cycles increasing, the fatigue damage extent increases, leading to the increase of fatigue life decreasing rate with occurrence cycle number of stochastic overloading stress.

Figure 5.5c–e shows the broken fiber faction versus applied cycle curves for different stochastic overloading stress levels and occurrence applied cycle number. The broken fiber fraction increases when stochastic overloading stress occurs, and the difference of broken fiber fraction between original peak stress and stochastic overloading stress level increases with applied cycle.

Table 5.5 shows the fatigue limit stress and broken fiber fraction at different occurrence applied cycle number and stochastic overloading stress. When stochastic overloading stress σ_s = 145 MPa occurs at applied cycle N_s = 10, 10^2, 10^3, 10^4, and 10^5, the broken fiber fraction increases from q = 0.00767, 0.01251, 0.03658, 0.07519, and 0.14038 under σ_{limit} = 143 MPa to q = 0.00834, 0.01359, 0.03969, 0.08145, and 0.15161; when stochastic overloading stress σ_s = 150 MPa occurs at applied cycle N_s = 10, 10^2, 10^3, 10^4, and 10^5, the broken fiber fraction increases to q = 0.01021, 0.01663, 0.04842, 0.09888, and 0.1825; and, when stochastic overloading stress σ_s = 155 MPa occurs at applied cycle N = 10, 10^2, 10^3, 10^4, and 10^5, the broken fiber fraction increases to q = 0.01241, 0.02021, 0.05864, 0.11905, and 0.21754.

5.3.5 3D C/SiC Composite

3D T-300™ carbon fiber-reinforced silicon carbide composite was fabricated using the chemical vapor infiltration (CVI) method. Low-pressure I-CVI was employed to deposit a pyrolytic carbon layer and the silicon carbide matrix. A thin pyrolytic carbon layer was deposited on the surface of the carbon fiber as the interfacial layer with C_4H_{10} at 950–1000 °C. The thickness of pyrolytic carbon layer was approximately 0.2 μm. The fiber volume is V_f = 0.4, and the average tensile strength is approximately σ_{uts} = 276 MPa. The dog-bone-shaped specimens were cut from composite panels using the diamond tooling, and then coated with a SiC coating. The tension–tension fatigue tests at room temperature were conducted on a servo hydraulic mechanical testing machine. The fatigue experiments were performed under load control at a loading frequency f = 60 Hz. The fatigue load ratio ($\sigma_{\text{min}}/\sigma_{\text{max}}$) was R = 0.1. The loading frequency affects the fatigue life and fatigue limit stress. At room temperature, when the loading frequency increases, the fatigue limit stress also increases.

Figure 5.6a shows the experimental and predicted fatigue life S–N curves of 3D C/SiC composite. When the fatigue limit applied cycle number is defined to be N_{limit} = 10^6, the corresponding predicted fatigue limit stress is approximately σ_{limit} = 236 MPa (approximately 85.5%σ_{uts}).

Figure 5.6b shows the fatigue life decreasing rate versus the occurrence cycle number of stochastic overloading curves for different stochastic overloading stress levels of σ_s = 240, 245, and 250 MPa (i.e., approximately 1.017, 1.038, and 1.059 fatigue limit stress). During the application of CMC components, the overloading stress level is not high, and the low overloading stress level is chosen for analysis. Under the same stochastic overloading stress level, the fatigue life decreasing rate increases with occurrence applied cycle of stochastic overloading. Under σ_s =

Table 5.5 Fatigue limit stress and broken fiber fraction of 2.5D C/SiC composite under stochastic overloading stress

$\sigma_{max}=$ 143 MPa	N_f		$N=1$	$N=10$	$N=10^2$	$N=10^3$	$N=10^4$	$N=10^5$
	1,012,346	q	0.00574	0.00767	0.01251	0.03658	0.07519	0.14038
$\sigma_s=$ 145 MPa $N=10$	N_f		$N=1$	$N=10$	$N=10^2$	$N=10^3$	$N=10^4$	$N=10^5$
	1,005,026	q	0.00574	0.00834	0.01317	0.03724	0.07585	0.14105
$\sigma_s=$ 145 MPa $N=10^2$	N_f		$N=1$	$N=10$	$N=10^2$	$N=10^3$	$N=10^4$	$N=10^5$
	1,000,463	q	0.00574	0.00767	0.01359	0.03766	0.07627	0.14146
$\sigma_s=$ 145 MPa $N=10^3$	N_f		$N=1$	$N=10$	$N=10^2$	$N=10^3$	$N=10^4$	$N=10^5$
	978,311	q	0.00574	0.00767	0.01251	0.03969	0.0783	0.1435
$\sigma_s=$ 145 MPa $N=10^4$	N_f		$N=1$	$N=10$	$N=10^2$	$N=10^3$	$N=10^4$	$N=10^5$
	944,705	q	0.00574	0.00767	0.01251	0.03658	0.08145	0.14664
$\sigma_s=$ 145 MPa $N=10^5$	N_f		$N=1$	$N=10$	$N=10^2$	$N=10^3$	$N=10^4$	$N=10^5$
	893,212	q	0.00574	0.00767	0.01251	0.03658	0.07519	0.15161
$\sigma_s=$ 150 MPa $N=10$	N_f		$N=1$	$N=10$	$N=10^2$	$N=10^3$	$N=10^4$	$N=10^5$
	984,589	q	0.00574	0.01021	0.01504	0.03911	0.07772	0.14292
$\sigma_s=$ 150 MPa $N=10^2$	N_f		$N=1$	$N=10$	$N=10^2$	$N=10^3$	$N=10^4$	$N=10^5$
	967,500	q	0.00574	0.00767	0.01663	0.0407	0.07931	0.1445
$\sigma_s=$ 150 MPa $N=10^3$	N_f		$N=1$	$N=10$	$N=10^2$	$N=10^3$	$N=10^4$	$N=10^5$
	886,901	q	0.00574	0.00767	0.01251	0.04842	0.08703	0.15223
$\sigma_s=$ 150 MPa $N=10^4$	N_f		$N=1$	$N=10$	$N=10^2$	$N=10^3$	$N=10^4$	$N=10^5$
	772,084	q	0.00574	0.00767	0.01251	0.03658	0.09888	0.16408
$\sigma_s=$ 150 MPa $N=10^5$	N_f		$N=1$	$N=10$	$N=10^2$	$N=10^3$	$N=10^4$	$N=10^5$
	613,579	q	0.00574	0.00767	0.01251	0.03658	0.07519	0.1825
$\sigma_s=$ 155 MPa $N=10$	N_f		$N=1$	$N=10$	$N=10^2$	$N=10^3$	$N=10^4$	$N=10^5$
	960,848	q	0.00574	0.01241	0.01725	0.04132	0.07993	0.14512
$\sigma_s=$ 155 MPa $N=10^2$	N_f		$N=1$	$N=10$	$N=10^2$	$N=10^3$	$N=10^4$	$N=10^5$
	929,613	q	0.00574	0.00767	0.02021	0.04427	0.08288	0.14808

(continued)

Table 5.5 (continued)

$\sigma_s =$ 155 MPa $N = 10^3$	N_f		$N = 1$	$N = 10$	$N = 10^2$	$N = 10^3$	$N = 10^4$	$N = 10^5$
	787,301	q	0.00574	0.00767	0.01251	0.05864	0.09725	0.16245
$\sigma_s =$ 155 MPa $N = 10^4$	N_f		$N = 1$	$N = 10$	$N = 10^2$	$N = 10^3$	$N = 10^4$	$N = 10^5$
	599,723	q	0.00574	0.00767	0.01251	0.03658	0.11905	0.18425
$\sigma_s =$ 155 MPa $N = 10^5$	N_f		$N = 1$	$N = 10$	$N = 10^2$	$N = 10^3$	$N = 10^4$	$N = 10^5$
	373,608	q	0.00574	0.00767	0.01251	0.03658	0.07519	0.21754

240 MPa, the fatigue life decreasing rate increases from $\Lambda = 0.22884$ at $N_s = 10$ to $\Lambda = 0.73423$ at $N_s = 10^5$; under $\sigma_s = 245$ MPa, the fatigue life decreasing rate increases from $\Lambda = 0.46292$ at $N_s = 10$ to $\Lambda = 0.94456$ at $N_s = 10^4$; and under $\sigma_s = 250$ MPa, the fatigue life decreasing rate increases from $\Lambda = 0.64272$ at $N_s = 10$ to $\Lambda = 0.98713$ at $N_s = 10^3$. When the occurrence applied cycle is between $N_s = 10$ and 10^2, the fatigue life decreasing rate increases rapidly; however, when the occurrence applied cycle is between $N_s = 10^2$ and 10^5, the fatigue life decreasing rate increases slowly. The occurrence of stochastic overloading stress at initial stage of cyclic fatigue loading deteriorates fatigue damage evolution, i.e., decreasing matrix crack spacing in transverse and longitudinal yarns, increasing interface debonding length, and broken fiber fraction; however, when matrix cracking and interface wear approach steady-state, the effect of stochastic overloading on fatigue damage or fatigue life decreasing rate decreases.

Figure 5.6c–e shows the broken fiber faction versus applied cycle curves for different stochastic overloading stress levels and occurrence cycle number. The broken fiber fraction increases when stochastic overloading stress occurs, and the difference of broken fiber fraction between original peak stress and stochastic overloading stress level increases with applied cycle.

Table 5.6 shows the fatigue limit stress and broken fiber fraction at different occurrence cycle number and stochastic overloading stress. When stochastic overloading stress $\sigma_s = 240$ MPa occurs at applied cycle $N_s = 10$, 10^2, 10^3, 10^4, and 10^5, the broken fiber fraction increases from $q = 0.04324, 0.11794, 0.18509, 0.2122$, and 0.24367 under $\sigma_{\text{limit}} = 236$ MPa to $q = 0.04771, 0.12961, 0.2026, 0.23189$, and 0.26575; when stochastic overloading stress $\sigma_s = 245$ MPa occurs at applied cycle $N_s = 10$, 10^2, 10^3, and 10^4, the broken fiber fraction increases to $q = 0.05383, 0.14538$, 0.22602, and 0.25812; and, when stochastic overloading stress $\sigma_s = 250$ MPa occurs at applied cycle $N = 10$, 10^2, and 10^3, the broken fiber fraction increases to $q =$ 0.06055, 0.1625, and 0.25116.

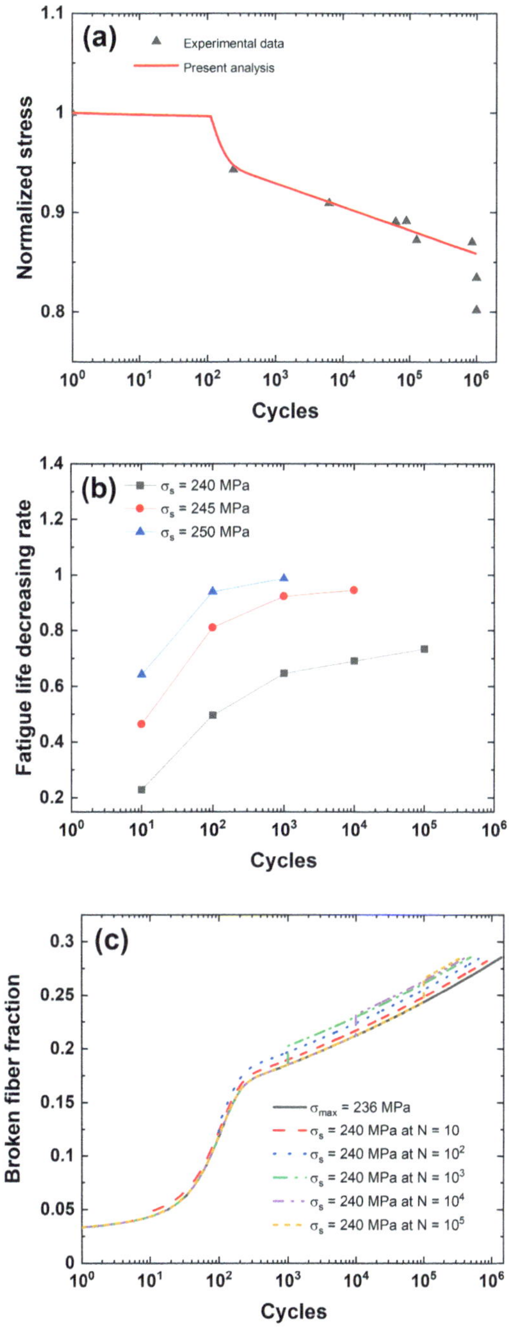

Fig. 5.6 **a** Experimental and predicted fatigue life S–N curves; **b** the fatigue life decreasing rate versus occurrence applied cycle curve for different stochastic overloading stress levels; **c** the broken fibers fraction versus applied cycle curves under $\sigma_{\text{limit}} = 236$ MPa and stochastic overloading stress of $\sigma_s = 240$ MPa at $N_s = 10, 10^2, 10^3, 10^4$, and 10^5; **d** the broken fibers fraction versus applied cycle curves under $\sigma_{\text{limit}} = 236$ MPa and stochastic overloading stress of $\sigma_s = 245$ MPa at $N_s = 10, 10^2, 10^3$, and 10^4; and **e** the broken fibers fraction versus applied cycle curves under $\sigma_{\text{limit}} = 236$ MPa and stochastic overloading stress of $\sigma_s = 250$ MPa at $N_s = 10, 10^2$, and 10^3 of 3D C/SiC composite

Fig. 5.6 (continued)

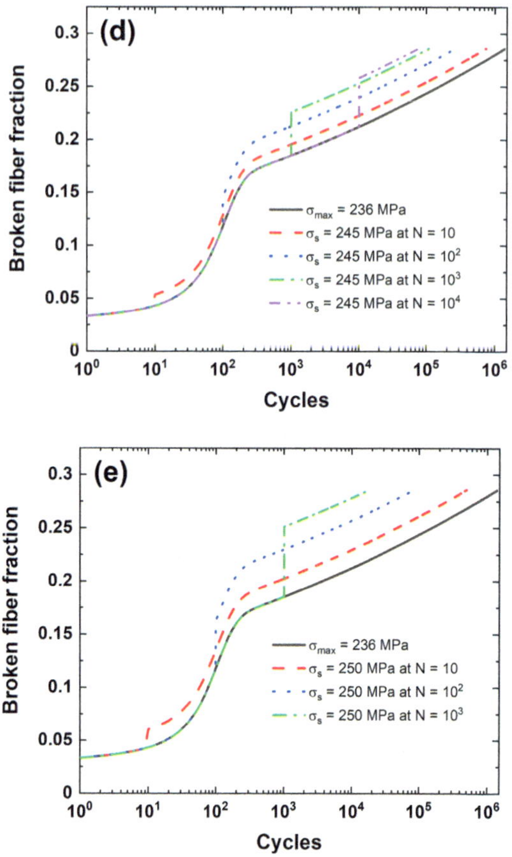

5.4 Discussion

Figure 5.7 shows the fatigue life decreasing rate versus stochastic overloading stress for different occurrence cycle number of different C/SiC composites. The fatigue life decreasing rate increases with stochastic overloading stress level for different fiber preforms (i.e., unidirectional, cross-ply, 2D, 2.5D, and 3D). However, with increasing applied cycles, the evolution of fatigue life decreasing rate with stochastic overloading stress depends on the fiber preforms, which indicates that the fiber preforms affect the fatigue damage evolution process.

For $N_s = 10$, 10^2, 10^3, and 10^4, under the same stochastic overloading stress level, the fatigue life decreasing rate is the highest for cross-ply C/SiC composite, which indicates that the fiber preform of cross-ply is very sensitive to the stochastic overloading stress.

For $N_s = 10$, 10^2, 10^3, 10^4, and 10^5, under the same stochastic overloading stress level, the fatigue life decreasing rate is the lowest for 2.5D C/SiC composite,

Table 5.6 Fatigue limit stress and broken fibers fraction of 3D C/SiC composite under stochastic overloading stress

$\sigma_{max} =$ 236 MPa	N_f		$N = 1$	$N = 10$	$N = 10^2$	$N = 10^3$	$N = 10^4$	$N = 10^5$
	1,383,192	q	0.03314	0.04324	0.11794	0.18509	0.2122	0.24367
$\sigma_s =$ 240 MPa $N = 10$	N_f		$N = 1$	$N = 10$	$N = 10^2$	$N = 10^3$	$N = 10^4$	$N = 10^5$
	1,066,666	q	0.03314	0.04771	0.12242	0.18957	0.21668	0.24815
$\sigma_s =$ 240 MPa $N = 10^2$	N_f		$N = 1$	$N = 10$	$N = 10^2$	$N = 10^3$	$N = 10^4$	$N = 10^5$
	696,321	q	0.03314	0.04324	0.12961	0.19676	0.22387	0.25534
$\sigma_s =$ 240 MPa $N = 10^3$	N_f		$N = 1$	$N = 10$	$N = 10^2$	$N = 10^3$	$N = 10^4$	$N = 10^5$
	488,262	q	0.03314	0.04324	0.11794	0.2026	0.22971	0.26118
$\sigma_s =$ 240 MPa $N = 10^4$	N_f		$N = 1$	$N = 10$	$N = 10^2$	$N = 10^3$	$N = 10^4$	$N = 10^5$
	426,789	q	0.03314	0.04324	0.11794	0.18509	0.23189	0.26336
$\sigma_s =$ 240 MPa $N = 10^5$	N_f		$N = 1$	$N = 10$	$N = 10^2$	$N = 10^3$	$N = 10^4$	$N = 10^5$
	367,609	q	0.03314	0.04324	0.11794	0.18509	0.2122	0.26575
$\sigma_s =$ 245 MPa $N = 10$	N_f		$N = 1$	$N = 10$	$N = 10^2$	$N = 10^3$	$N = 10^4$	$N = 10^5$
	742,891	q	0.03314	0.05383	0.12853	0.19568	0.22279	0.25426
$\sigma_s =$ 245 MPa $N = 10^2$	N_f		$N = 1$	$N = 10$	$N = 10^2$	$N = 10^3$	$N = 10^4$	$N = 10^5$
	262,035	q	0.03314	0.04324	0.14538	0.21253	0.23963	0.2711
$\sigma_s =$ 245 MPa $N = 10^3$	N_f		$N = 1$	$N = 10$	$N = 10^2$	$N = 10^3$	$N = 10^4$	$N = 10^5$
	107,854	q	0.03314	0.04324	0.11794	0.22602	0.25313	0.2846
$\sigma_s =$ 245 MPa $N = 10^4$	N_f		$N = 1$	$N = 10$	$N = 10^2$	$N = 10^3$	$N = 10^4$	$N = 76{,}689$
	76,689	q	0.03314	0.04324	0.11794	0.18509	0.25812	0.28571
$\sigma_s =$ 250 MPa $N = 10$	N_f		$N = 1$	$N = 10$	$N = 10^2$	$N = 10^3$	$N = 10^4$	$N = 10^5$
	494,186	q	0.03314	0.06055	0.13525	0.2024	0.22951	0.26098
$\sigma_s =$ 250 MPa $N = 10^2$	N_f		$N = 1$	$N = 10$	$N = 10^2$	$N = 10^3$	$N = 10^4$	$N = 84{,}191$
	84,191	q	0.03314	0.04324	0.1625	0.22966	0.25676	0.28571
$\sigma_s =$ 250 MPa $N = 10^3$	N_f		$N = 1$	$N = 10$	$N = 10^2$	$N = 10^3$	$N = 10^4$	$N = 17{,}798$
	17,798	q	0.03314	0.04324	0.11794	0.25116	0.27827	0.28571

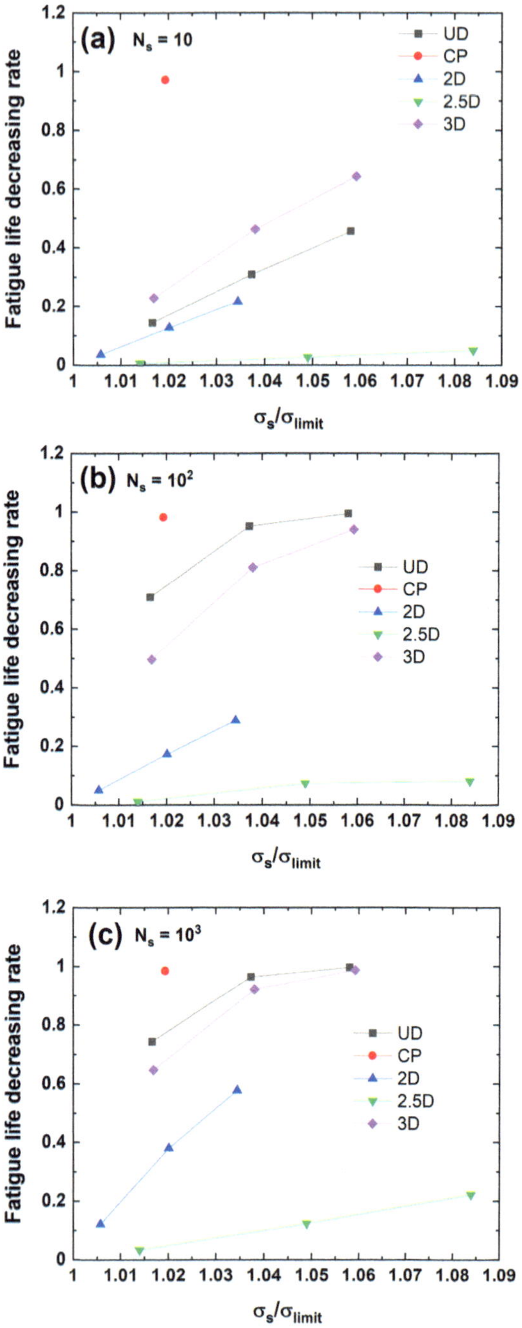

Fig. 5.7 Fatigue life decreasing rate versus stochastic overloading stress for different occurrence applied cycle of **a** $N_s = 10$; **b** $N_s = 10^2$; **c** $N_s = 10^3$; **d** $N_s = 10^4$; and **e** $N_s = 10^5$

Fig. 5.7 (continued)

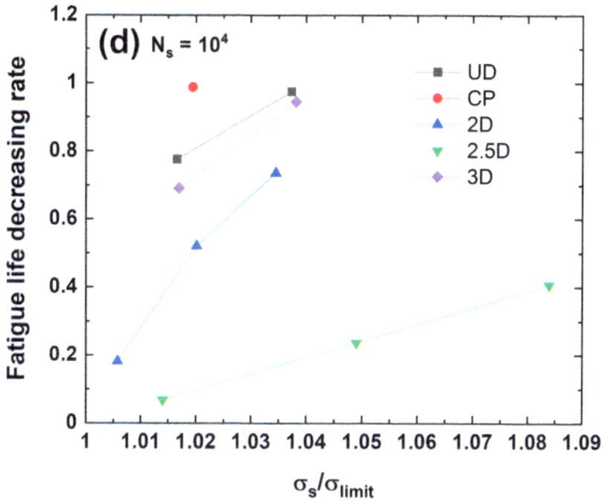

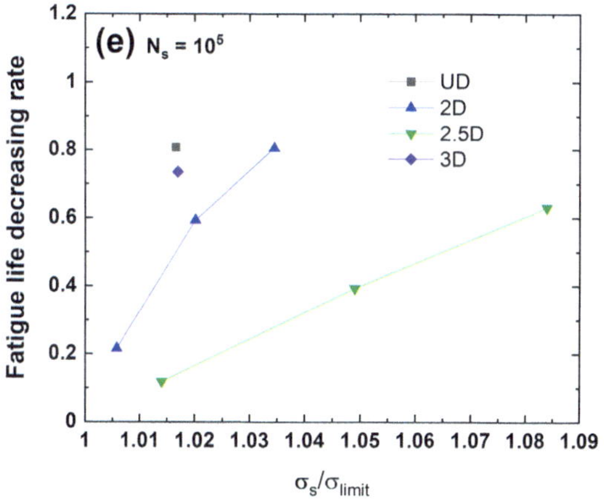

which indicates that the fiber preform of 2.5D has high resistance to the stochastic overloading stress.

Among UD, 2D, and 3D C/SiC composites, at the initial stage of cyclic fatigue loading, i.e., $N_s = 10$, under the same stochastic overloading stress, the fatigue life decrease rate of 3D C/SiC is the highest; however, with increasing applied cycle number, the fatigue life decreasing rate of UD C/SiC composite is the highest under the same stochastic overloading stress.

5.5 Conclusion

In this chapter, a micromechanical fatigue limit stress model of fiber-reinforced CMCs subjected to stochastic overloading stress is developed. Fatigue limit stress for different C/SiC composites is predicted. Relationships between fatigue life decreasing rate, stochastic overloading stress, corresponding occurrence cycle number, and broken fiber fraction are analyzed.

Under the same stochastic overloading stress level, the fatigue life decreasing rate increases with stochastic overloading stress level and occurrence applied cycle of stochastic overloading for different fiber preforms. The broken fiber fraction increases when stochastic overloading stress occurs, and the difference of broken fiber fraction between fatigue limit stress and stochastic overloading stress level increases with applied cycle.

Under the same stochastic overloading stress level and occurrence applied cycle, the fatigue life decreasing rate is the highest for cross-ply C/SiC composite, and lowest for 2.5D C/SiC composite.

For UD, CP, and 3D C/SiC composites, when applied cycle is between $N_s = 10$ and 10^2, the fatigue life decreasing rate increases rapidly with occurrence applied cycle; however, when applied cycle is higher than $N_s = 10^2$, the fatigue life decreasing rate increases slowly with applied cycle.

For 2D and 2.5D C/SiC composites, when applied cycle is between $N_s = 10$ and 10^2, the fatigue life decreasing rate increases slowly with occurrence applied cycle; however, when applied cycle is between $N_s = 10^2$ and 10^5, the fatigue life decreasing rate increases rapidly with applied cycle.

References

Curtin WA (1991) Theory of mechanical properties of ceramic-matrix composites. J Am Ceram Soc 74:2837–2845. https://doi.org/10.1111/j.1151-2916.1991.tb06852.x

Du SM, Qiao SR, Ji GC, Han D (2009) Tension-tension fatigue behavior of 3D-C/SiC composite at room temperature and 1300 °C. Mater Eng 9:22–25

Evans AG (1997) Design and life prediction issues for high-temperature engineering ceramics and their composites. Acta Mater 45:23–40. https://doi.org/10.1016/S1359-6454(96)00143-7

Li LB (2015a) Tension-tension fatigue behavior of unidirectional C/SiC ceramic-matrix composite at room and elevated temperature and 800 °C in air atmosphere. Materials 8:3316–3333. https://doi.org/10.3390/ma8063316

Li LB (2015b) Damage evolution and life prediction of cross-ply C/SiC ceramic-matrix composite under cyclic fatigue loading at room temperature and 800 °C in air. Materials 8:8539–8560. https://doi.org/10.3390/ma8125474

Li LB (2015c) Fatigue life prediction of carbon fiber-reinforced ceramic-matrix composites at room and elevated temperatures. Part II: experimental comparisons. Appl Compos Mater 22:961–972. https://doi.org/10.1007/s10443-015-9445-4

Li LB (2016a) Fatigue life prediction of carbon fiber-reinforced ceramic-matrix composites at room and elevated temperatures. Part I: experimental analysis. Appl Compos Mater 23:101–117. https://doi.org/10.1007/s10443-015-9446-3

Li LB (2016b) Fatigue life prediction of fiber-reinforced ceramic-matrix composites with different fiber preforms at room and elevated temperatures. Materials 9:207. https://doi.org/10.3390/ma9030207

Li LB (2019a) Thermomechanical fatigue of ceramic-matrix composites. Wiley-VCH, Weinheim

Li LB (2019b) Stress-rupture of fiber-reinforced ceramic-matrix composites with stochastic loading at intermediate temperatures. Part I: theoretical analysis. Materials 12:3123. https://doi.org/10.3390/ma12193123

Li LB (2020a) Time-dependent mechanical behavior of ceramic-matrix composites at elevated temperatures. Springer Nature Singapore Pte Ltd., Singapore. https://doi.org/10.1007/978-981-15-3274-0

Li LB (2020b) A micromechanical fatigue limit stress model of fiber-reinforced ceramic-matrix composites under stochastic overloading stress. Materials 13:3304. https://doi.org/10.3390/ma13153304

Li LB, Reynaud P, Fantozzi G (2020) Cyclic-dependent damage evolution in self-healing woven SiC/[Si-B-C] ceramic-matrix composites at elevated temperatures. Materials 13:1478. https://doi.org/10.3390/ma13061478

Naslain R (2004) Design, preparation and properties of non-oxide CMCs for application in engines and nuclear reactors: an overview. Compos Sci Technol 64:155–170. https://doi.org/10.1016/S0266-3538(03)00230-6

Reynaud P (1996) Cyclic fatigue of ceramic-matrix composites at ambient and elevated temperatures. Compos Sci Technol 56:809–814

Ruggles-Wrenn MB, Lee MD (2016) Fatigue behavior of an advanced SiC/SiC ceramic composite with a self-healing matrix at 1300 °C in air and in steam. Mater Sci Eng A 677:438–445. https://doi.org/10.1016/j.msea.2016.09.076

Ruggles-Wrenn MB, Sharma V (2011) Effects of steam environment on fatigue behavior of two SiC/[SiC + Si3N4] ceramic composites at 1300 °C. Appl Compos Mater 18:385–396. https://doi.org/10.1007/s10443-010-9163-x

Ruggles-Wrenn MB, Christensen DT, Chamberlain AL, Lane JE, Cook TS (2011) Effect of frequency and environment on fatigue behavior of a CVI SiC/SiC ceramic matrix composite at 1200 °C. Compos Sci Technol 71:190–196. https://doi.org/10.1016/j.compscitech.2010.11.008

Ruggles-Wrenn MB, Boucher N, Przybyla C (2018) Fatigue of three advanced SiC/SiC ceramic matrix composites at 1200 °C in air and in steam. Int J Appl Ceram Technol 15:3–15. https://doi.org/10.1111/ijac.12773

Shuler SF, Holmes JW, Wu X, Roach D (1993) Influence of loading frequency on the room-temperature fatigue of a carbon-fiber/SiC-matrix composite. J Am Ceram Soc 76:2327–2336. https://doi.org/10.1111/j.1151-2916.1993.tb07772.x

Suzuki T, Mahfuz H, Takanashi M (2019) A new stiffness degradation model for fatigue life prediction of GFRPs under random loading. Int J Fatigue 119:220–228. https://doi.org/10.1016/j.ijfatigue.2018.09.021

Yang F (2011) Research on fatigue behavior of 2.5d woven ceramic matrix composites. Master Thesis, Nanjing: Nanjing University of Aeronautics and Astronautics

Zhu S, Mizuno M, Kagawa Y, Mutoh Y (1999) Monotonic tension, fatigue and creep behavior of SiC-fiber-reinforced SiC-matrix composites: a review. Compos Sci Technol 59:833–851. https://doi.org/10.1016/S0266-3538(99)00014-7

Printed by Printforce, the Netherlands